Praise from the Experts

"Using nontechnical terms, this book skillfully guides the user through the p..........SAS Enterprise Miner software to answer critical questions such as 'What are the characteristics of my best customers?', 'How can I optimize their value?', and 'How can I use this knowledge to grow my business?'. The ease and precision with which these techniques are explained will be appreciated by marketers, managers, analysts, and anyone else whose role it is to turn customer data into actionable knowledge."

> **C. Olivia Parr Rud, M.S.**
> **Data Mining Strategist/Author/Facilitator**
> **OLIVIAGroup**

"I am really looking forward to using this book in a data mining course. The exposition with illustrations on real data sets is terrific. There is a fine recognition of the importance of data preparation, as well. Excellent!"

> **Mark E. Johnson, Fellow ASA**
> **Professor of Statistics and**
> **Interim Director of the Data Mining Program**
> **University of Central Florida**

"The application of SAS Enterprise Miner and data mining analytics continues to broaden to new domains, from medical epidemiology to advanced business practices. Randy Collica provides us with an intuitive, hands-on guide to implementing the science behind comprehensively and accurately understanding one's customers. The efficiency of these CRM techniques has been maximized through the described segmentation and clustering methods and through the mining of underutilized data types, such as free text. These techniques are presented in a detailed, yet appealing and sensible manner; one that is a welcomed contribution to the state of the art in data mining analytics."

> **Daniel C. Payne, Ph.D., M.S.P.H**
> **Epidemiologist**

"*CRM Segmentation and Clustering Using SAS® Enterprise Miner™* is an excellent how-to guide to segmentation and clustering for SAS Enterprise Miner users. While it serves as a solid introductory text for the data mining novice, it also serves as a reference guide for experienced data miners.

"Overall, this book contains handy step-by-step instructions for implementing these techniques in SAS Enterprise Miner. It explains how to handle, real-life, non-normally distributed data and provides literally dozens of examples from retail to banking to the experimental sciences and more."

> **Tony Lee, Ph.D.**
> **Senior Manager, Customer Insight and Analysis**
> **PetSmart**

CRM Segmentation and Clustering
Using **SAS**® **Enterprise Miner**™

Randall S. Collica

The correct bibliographic citation for this manual is as follows: Collica, Randall S. 2007. *CRM Segmentation and Clustering Using SAS® Enterprise Miner™*. Cary, NC: SAS Institute Inc.

CRM Segmentation and Clustering Using SAS® Enterprise Miner™

This book is dedicated to my lovely wife Nanci and
to my children, Janelle, Brian, Danae, Jamie, and Carmella.

iv

Contents

Foreword

Artists call it white shock—the paralyzing effect brought on by a blank canvas. Where should the first brush stroke be applied? Every student who has ever taken a class in some technical topic and then gone back to the office to try out the newly learned techniques is familiar with the feeling. You thought you understood everything the teacher was saying, but now you are staring at a blank pad of paper, or the open program editor window on your computer screen, with no idea how to get started.

If the project you are embarking on involves customer segmentation or clustering, then Randy Collica has written the cure for your white shock. This book is not so much a discussion of segmentation and clustering as it is a tutorial on segmentation and clustering; it tells you what to do step by step.

There is a large, but sometimes unacknowledged, difference between learning about things and learning to do things. You can learn a lot about sailing or painting or dancing by attending lectures or reading a book. When you put the book down and pick up the tiller, paintbrush, or dancing shoes, you will not leave a straight wake behind the boat, capture the play of light on water on your canvas, or stay on beat dancing a mambo. Learning technical skills is no different. You need to understand the business context to know what you want to do. You need strong theoretical background to understand why what you are doing might work. And, you need practice to actually learn how to do it. Working through the exercises in this book will give you that practice.

If practice is so important to learning, why aren't more books written this way? As an author myself, I think I know. We writers would generally like to reach as broad an audience as possible so we can sell more copies of our books. As soon as we leave generalities behind, we have to make choices that threaten to narrow our audience. Concrete examples necessarily come from particular industries or areas of study. Step-by-step instructions must assume the availability of particular software packages. Randy Collica is able to give detailed instructions because he has made the choice to assume the reader has access to SAS Enterprise Miner, a very complete set of data mining tools that sits on top of and integrates with the SAS programming environment. This choice will doubtless deny the book some readers, but it enables the tutorial approach.

Another important tool for learning is realistic data. This point is worth saying more about. Unrealistic data sets lead to unrealistic results. This is frustrating to the student. In real life, the more you know about the research domain or business context, the better your data mining results will be. Subject matter expertise gives you a head start. You know what variables ought to be predictive and have good ideas about new ones to derive. Fake data does not reward these good ideas because patterns that should be in the data are missing and patterns that shouldn't be there have been introduced inadvertently. Real data is hard to come by, not least because real data may reveal more than its owners are willing to share about their business operations. As a result, many instructors make do with artificially constructed data sets. The examples and exercises in this book make use of realistic data that is supplied with the book on a CD along with the SAS code used in the examples.

As it happens, I know the back story of one of the data sets used in this book. Several years ago, my company, Data Miners, did a clustering project for *The Boston Globe*. This involved taking census data for scores of towns in eastern Massachusetts and southern New Hampshire and clustering them into groups with similar demographics. Among other things, we used this to study household penetration (percentage of households subscribing to the Globe) and how it varied by cluster. When Gordon Linoff and I set out to write a data mining class for the SAS Business Knowledge Series, we thought that project would make a good example. We did not want to make use of our client's potentially sensitive circulation data for this exercise, so we substituted penetration figures derived from publicly available census bureau data—namely, the percentage of homes heated primarily by wood. For good measure, we moved the study from Massachusetts to New York. In our class, we did not use the data for a clustering exercise. Instead, we used it for building regression models with product penetration as the dependent variables. When Randy Collica asked if he could use our course data set for this book, we readily agreed (after all, the data is all publicly available anyway). I was surprised and amused to see that the data has come full circle; it is once again used for clustering, just as it was in the original project for *The Boston Globe*.

So, fire up SAS Enterprise Miner, insert the CD, and get to work!

Michael J. A. Berry
Data Miners Inc.

Preface

Introduction

This book aims at one of the basic beginning points when initiating a Customer Relationship Management (CRM) program: understanding your customers. Unless you understand your customers, the relationship part of CRM is almost entirely absent. Those who want to "know" your customers using analytical CRM techniques will value the applications presented in this book.

You do not necessarily need a formal background in statistics because in SAS Enterprise Miner much of what you need is contained in the software package, however, additional SAS code and macros are provided for enhanced capability are on the CD that accompanies this book. (See the appendixes for a description of the SAS data sets, SAS programs and macros, and SAS Enterprise Miner templates used in the examples.) A rudimentary understanding of data mining techniques is helpful but not mandatory. Also, I recommend that you read the introductory material in SAS Enterprise Miner documentation so that you will have an elementary understanding of how data mining projects and process flow diagrams are created and managed.

This book could be used as a companion course introducing data mining applications in information sciences, computer science, or marketing information management. Detailed algorithms are not developed in this book; however, many references are made to recent literature for further reading.

The number of books and journal literature in the field of data mining has increased greatly in the past several years. Most books tend to focus on the algorithmic nature of data mining, and some, like Dorian Pyle's *Business Modeling and Data Mining*, focus on data preparation. In this book, I show you how to use the most commonly available techniques and how to branch out into some new ones such as text mining, which is covered in the last chapter. I show you how to perform these techniques using SAS Enterprise Miner software and how to use them in the context of CRM. I endeavored to make this a how-to book for segmentation and clustering rather than a theoretical one. I do review some of the basic equations that will help you understand topics; however, I give no formal proofs. References are given at the end of each chapter, where applicable, along with some suggested readings. In a few chapters, additional exercises are also provided to help you develop the concepts further.

Even though the context is customer analyses, you can use these concepts in other fields such as medical diagnosis, insurance claims, fraud detection, and others. Segmenting your customers for more intelligent use and getting closer to the one-to-one customer relationship is what most businesses now want to achieve.

This book focuses less on actual theory and more on how to implement existing theory in the actual practice of segmentation using data mining techniques. When I started doing CRM segmentation work in the late 1990s, I desired to have a segmentation guide that I could use to help me implement the techniques that I read about. Techniques such as clustering, decision trees, regressions, neural networks, and the like, are well-documented. However, I found that though many texts describe the algorithms well, very little is mentioned on how to use these techniques in practice. At the end of each chapter in this book, I offer references for further reading. Many of these texts are excellent at describing the techniques algorithmically, and some contain business cases as well. I hope that this book will help you in your data mining endeavors as much as writing it helped me.

Overview of Chapters

This book is broken down into three parts, each of which increases in complexity. **Part 1, "The Basics,"** discusses the basics in terms of what segmentation is comprised of, and measures of distance and association. **Part 2, "Segmentation Galore,"** dives right into the core of segmentation using RFM cells and moves into other techniques such as clustering. **Part 3, "Beyond Traditional Segmentation,"** reviews some advanced techniques for segmentation, such as how to segment customers based on their product affinity, and discusses some of the measures of product affinity as well as some of the pitfalls.

Part 1, The Basics

Chapter 1, "Introduction," introduces the basic concept of segmentation in light of CRM and defines some of the techniques used to achieve segmentation of your customer database records.

Chapter 2, "Why Segment? The Motivation for Segment-based Descriptive Models," presents the motivation for customer segmentation and the concept of *descriptive* versus *predictive* models. This chapter discusses why you would want to classify or group customers or prospects into various segments and how to use them. The data assay and profile are reviewed, as well as how these can be used to understand your data prior to mining.

Chapter 3, "Distance: The Basic Measures of Similarity and Association," describes how to measure distance from one customer record to another and also introduces the measure of association. These concepts are key to understanding what types of settings are needed in the various techniques used, such as clustering, decision trees, and memory-based reasoning, which are discussed in later chapters.

Part 2, Segmentation Galore

Chapter 4, "Segmentation Using a Cell-based Approach," introduces the concepts of recency, frequency, and monetary value (RFM) and discusses how to compute these cells and score your customers for each of the cell groups.

Chapter 5, "Segmentation of Several Attributes with Clustering," introduces segmentation with the use of clustering algorithms on a few customer attributes. An example that involves 100,000 customers demonstrates the concept and shows the detail of creating the process flow diagram in SAS Enterprise Miner. This chapter also discusses the default coding of categorical and binary versus ordinal variables and shows how these settings can produce different results.

Chapter 6, "Clustering of Many Attributes," extends the clustering techniques when many customer attributes are being used. A new example that has a fairly large set of variables is introduced, and it shows how you might attack this problem with some pre-processing prior to performing clustering.

Chapter 7, "When and How to Update Cluster Segments," presents several practical issues that arise after you start using cluster segments. This chapter discusses model shelf-life, or the practical usable life of a model before it needs to be refitted. You will learn how to tell that the cluster segments have "moved" from their original model when your input data has been refreshed.

Chapter 8, "Using Segments in Predictive Models," breaks away from the topic of pure segmentation and discusses how the segments can be used to partition the data space and, in so doing, reduce the dimensionality of the data. It is now somewhat easier to generate a predictive model using the data. An example demonstrates a cluster segmentation and a predictive model to predict one cluster from the cluster analysis.

Part 3, Beyond Traditional Segmentation

Chapter 9, "Clustering and the Issue of Missing Data," reviews how missing data elements can affect data mining models, especially focusing on clustering. There are several methods for treating missing data in the cluster algorithm and also external and prior to clustering. The implementation and use of the data imputation node as well as the MI procedure is reviewed.

Chapter 10, "Product Affinity and Clustering of Product Affinities," shows you how, once segments are created, to estimate the affinity of products by transposing product transaction quantity data onto the customer data records that are segmented and thus estimate the affinity of products for each segment. In addition, this chapter describes how you can cluster the product affinities into various cluster segments. This chapter also reviews how to use product affinities within customer segments and how that knowledge can aid in the CRM learning process.

Chapter 11, "Computing Segments Using SOM/Kohonen for Clustering," introduces a special-purpose neural network called a self-organizing map (SOM) to cluster customer data. This type of clustering uses a neural network algorithm that can accept a large number of inputs and will cluster each record into a two-dimensional map of desired size.

Chapter 12, "Segmentation of Textual Data," introduces the exciting and relatively new technique of text mining and the concept of similarity, or association, is revisited. This chapter requires that you have SAS Text Miner, which is an add-on product to SAS Enterprise Miner. Although this topic could be a book in and of itself, the same basic concepts of clustering documents and combining this new information with the previous techniques makes this a powerful method for business intelligence applications and CRM in general.

Acknowledgments

There are a number of people who greatly helped make this book a reality. First, to my wife, Nanci, who never complained about my time spent in writing this book and, therefore, demonstrated great patience. In addition, my direct management at Hewlett-Packard Co., Sally Dyer and Carrie Fraser, encouraged me personally. I also must mention Patsy Poole, my acquisitions editor, who coordinated all the reviews, editing, and production logistics. I would also like to thank the following individuals who assisted so tirelessly in the review of this manuscript. This manuscript would not have been possible without the assistance of these people at SAS Institute: Ross Bettinger, Brent Cohen, Mary Crissey, Barry de Ville, Craig DeVault, Manya Mayes, Annette Sanders, Wayne Thompson, and Sue Walsh. Also, many thanks to Dr. Tom Bohannon, Dr. Goutam Chakraborty, and Dr. Joydeep Ghosh, professors who provided additional technical reviews.

To SAS Institute, who made this book possible, and their software SAS Enterprise Miner and SAS Text Miner. Their software, in my opinion, is one of the best commercially available data mining packages that exist today.

How to Use This Book

Each of the many examples in this book begins with a process flow table that outlines the steps that are necessary to complete the exercise. This process flow table gives the step number, the step description, and a brief rationale. The step detail is a statement outlining what is taking place in the overall data mining process flow. These individual steps are indicated in the exercise as **Step 1**, **Step 2**, and so on. Armed with the process flow table, the steps, and the snapshots of SAS Enterprise Miner process flow diagrams and intermediate steps, you should be able to navigate through an exercise with greater ease. It is my hope and desire that this book allows you to know your customers better and to gain insight by using SAS Enterprise Miner in a data-driven purposeful fashion.

P a r t 1

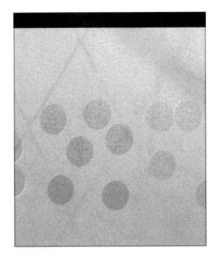

The Basics

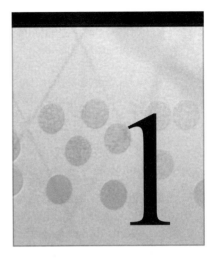

Introduction

1.1 What Is Segmentation in the Context of CRM?

Segmentation is in essence the process by which items or subjects are categorized or classified into groups that share similar characteristics. That characteristic could be one or more attributes. Segmentation also can be defined as subdividing the population according to already known *good discriminators*. Hand, Mannila, and Smyth distinguish between segmentation and clustering based on differing objectives (Hand, Mannila, and Smith 2001, p. 293). The terminologies used in clustering algorithms arose from various multiple disciplines such as computer science, biology, social science, statistics, and astronomy. Therefore, it is sometimes difficult to grasp the concepts in clustering with such widely varying terminology and syntax. In segmentation, the aim is simply to partition the data in a way that is convenient. Convenient may refer to something that is useful, as in marketing, for example. In clustering, the objective is to see if a sample of data is composed of natural subclasses or groups. This may be the objective in customer profiling. The analytical techniques involved in both of these objectives could very well be the same. There are a great number of methods and algorithms used in cluster analysis. The important thing is to match the method with your business objective as close as possible. This book's aim is to help you choose the method depending on your objective and to avoid mishaps in the analysis and interpretation. It is also to help you understand how to apply and implement these techniques using SAS Enterprise Miner. In Customer Relationship Management (CRM), segmentation is used to classify customers according to some similarity, such as industry, for example. This book describes the methods used to segment records in a database of customers; it is the how to of segmentation analysis.

If you can remember back in elementary school when selecting teams for softball or kickball, the team captains would always choose the tallest or strongest players first to be on the team, leaving the shortest to be last. The elementary school teacher would instead have everyone line up and call out numbers from one to four and then repeat so that each number that was the same would then be members of the same team. This was a form of undirected *segmentation* until the children caught on and tried to line up their friends to circumvent their teacher's method. The measure of similarity of the members was nothing more than the matching numbers assigned during the lineup. If the similarity were the height of the members, then after measuring the height of each individual, each would be sorted into teams according to each other's height, thus giving segments of members that have similar height. The characteristic of the segments then is strongly dependent on the measure of similarity used for each subject.

To apply this simple concept of similarity to a situation involving CRM, take for example, a marketing analyst who desires to segment his prospects into groups of industry segments. The analyst believes that marketing differently to each industry segment would produce a higher response and generate more revenue than not using any industry affiliation. In order to accomplish this task he records in his business-to-business (B-to-B) database each prospect's standard industry classification (SIC) code and then categorizes them according to the first two digits. This allows him to find the major industries in his database. The measure of similarity is the SIC code according to the government's coding of their primary business industry classification. This is now a segmentation of industry groups as illustrated in Table 1.1.

Table 1.1 Example of B-to-B Industry Segmentation

Record No	Prospect Company Name	SIC Code (2-digit)	SIC Description	Industry Segment
1	ABC Gravel and Sand Co.	14	Construction Sand and Gravel	Forest, Mining, and Metals
2	Metro Cable TV	48	Cable Television	Telecommunications
3	Joe's Computer Shop and Service	73	Computer Maintenance and Repair	Professional Services

Let's take another example. Owners of credit cards can be divided into subgroups according to how they use their card, what kind of items they purchase, how much money they spend, how often they use their card, and so on. It will be very useful for CRM purposes in marketing to identify the groups to which a card owner belongs, since he or she can then be targeted with special promotional material that might be of interest (and this clearly benefits the owner of the card as well as the card company). Look for further discussion on the benefits of why this might be so in Section 1.3.

In addition to spending patterns, purchase frequency, and so on, one can segment by any attribute recorded in a database. When multiple attributes are chosen, several problems arise in the computations that may be used to create the segments or clusters. For example, how does one choose a measurement scheme so that all characteristics are being measured on a similar scale? How can you determine the importance of the effect of each variable on the segment clusters? Issues like these will be discussed in later chapters, especially Chapters 3 through 6.

1.2 Types of Segmentation and Methods

There are many techniques for classifying records or rows in a database. For the purpose of this book, I will interchange the term segmentation with the phrase *record classification*, because in the context of CRM these can be used synonymously. In the world of computer science, there is a definite distinction between classification of records in a database and grouping or clustering records according to some criteria of similarity or likeness. Classification is typically referred to as assigning a record to one of a number of predetermined classes. Clustering is a set of algorithms used to partition records in a database according to a measure of similarity, and the number of cluster segments is not predetermined before the algorithm is applied to the database. This distinction becomes less important in business applications; however, it is useful to keep these definitions in mind. In order to discuss the types and uses of segmentation one needs to review the various capabilities that each type has to offer. What follows is only a partial list of the many types of segmentations that exist, but this should be useful for determining which set of techniques you may need to perform for solving the business problems at hand.

1.2.1 Customer Profiling

In profiling a set of customers, the typical reason for performing this analysis is to gain insight or an understanding of the four Ws—the who, what, where, and when of your customer base. A fifth W of why can also be added; however, the why is always a much more difficult customer attribute to collect. A typical business problem might involve a question from your field sales force like the following: I need to understand my customer base in the northwest area so I can deploy my field sales force accordingly. This kind of business question would require one to know how many customers exist in the northwest area as well as their recent purchases, what industries they mainly come from, and so on. A customer profile by geographic region will then help the business manager requesting the analysis to align the sales force with customers to achieve greater sales coverage and effectiveness in their customer base. The techniques used in this kind of profiling may include counting the number of customers by region or zip code range for each industry group or perhaps counting the number of customers who have made purchases within the last year and ones who have not. This can be a simple query to the customer database, but if the number of attributes desired is large, it may be an impossible database query and you will need to resort to a clustering algorithm. An example of a customer profile might look like the following two query results.

Table 1.2 Example of Customer Profiling in NW US Region (Profile by state)

Northwest Customer Sales by State

State	Total Sales	No of Customers
ID	$2,799,607	135
MO	$16,570,851	305
OR	$8,746,203	326
WA	$38,885,342	466

Table 1.3 Example of Customer Profiling in NW US Region (Profile by Major Metro/3 digit Postal Code – Only top 8 rows shown)

Northwest Customer Sales by Major Metro or 3digit Zip

Major Metro/Zip Code	Total Sales	No of Customers
SEA	$25,578,204	283
PDX	$3,971,539	172
STL	$3,562,242	91
974	$2,029,223	55
MKC	$1,412,478	55
982	$3,019,754	31
977	$841,117	31
834	$438,883	28

In essence, these reports from Tables 1.2 and 1.3 are *results* of segmentation. (E.g., the segments include the state as one segment, and the 3-digit postal abbreviation/major metro area code combined as another segment.) In this case, when there is no major metro code, the code abbreviation is used in its place. Then the sales and number of customers are aggregated (summed in this case) by each of these segments. This type of segmentation profiling will be discussed in further detail in Chapters 5 and 6. The output in Tables 1.2 and 1.3 were performed using HTML output settings on PC SAS with the output selected to create HTML with default settings. If this code were run in SAS Enterprise Miner, the output would be within the SAS Code output window. There will be more on the discussion of using SAS Code nodes in the examples in Chapter 2, "Why Segment? The Motivation for Segment-based Descriptive Models," and later chapters.

SAS Code to Generate Output in Tables 1.2 and 1.3

```
libname chapt1 'f:\chapter 1';   /* where F: is referring to your CD
drive */

data work.northwest;
  set chapt1.northwest;
  if majmet=' ' then majmet=substr(zip,1,3);
run;

proc summary data=work.northwest nway sum;
  class majmet state branch_code;
  var sales ;
  output out=work.nw_sum sum= ;
run;

title 'Northwest Customer Sales by State';
proc sql;
 select state label='State',
        sum(sales) as sum_sales label='Total Sales' format=dollar12.,
         sum(_freq_) as count label='No of Customers'
   from work.nw_sum as q1
 group by state
  ;
quit; title;

title 'Northwest Customer Sales by Major Metro or 3digit Zip';
proc sql;
 select majmet label='Major Metro/Zip Code',
        sum(sales) as sum_sales label='Total Sales' format=dollar12.,
         sum(_freq_) as count label='No of Customers'
   from work.nw_sum as q1
 group by majmet
   order by count descending;
quit; title;
```

1.2.2 Customer Likeness Clustering

A chain store or franchise might want to study whether its outlets are similar in terms of social neighborhood, size, staff numbers, vicinity to other shops, and so on. Their objective is to see if they have similar turnovers and yield similar revenues or profits. A beginning point might be to cluster the outlets in terms of these variables and to examine the distributions of turnovers and profits within each group. Another method would be to cluster just the turnovers and revenue/profit variables and then profile other variables of interest like geography, social neighborhood, etc. We will discuss methods of clustering in Chapters 5 and 6 and review some practical techniques of how and when to update those models for continued maintenance. A simple example using two varieties of orange sales (ORANGES sample data set from SAS Institute) produces the analysis of sales comparisons between six stores. Figure 1.1 shows the normalized distances of the six clusters from sales of two varieties of orange sales on six days of sales from six stores. The normalized distances are the results from clustering two types of orange sales data using the clustering node in SAS Enterprise Miner; the distance plot is the result of the MDS procedure. We will review this type of analysis in greater detail in Chapter 5, "Segmentation of Several Attributes with Clustering."

Figure 1.1 Orange Sales Clusters – Distance Plot from SAS Enterprise Miner

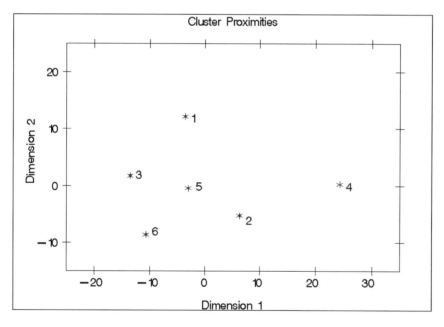

1.2.3 RFM Cell Classification Grouping

RFM stands for recency, frequency, and monetary value. *Recency* (a term typically used in direct marketing industry) is a measure of the time lag since your customer has either communicated or purchased last from your business. Recency can be measured in weeks, months, quarters, fiscal years, etc. *Frequency* is the quantity or volume of items or services purchased and can be single units or perhaps aggregated in deciles or whatever meaningful grouping. *Monetary value* is just that, a numeric currency figure representing the value of each of the frequency units or aggregated units that were purchased. RFM cells can be easily thought of in three dimensions as shown in Figure 1.2. Each customer will be classified into only one of the cells as the classification is applied to the customer database. We will be discussing this type of segmentation method and its uses in Chapter 4, "Segmentation Using a Cell-based Approach."

Figure 1.2 RFM Cell Pictorial Description

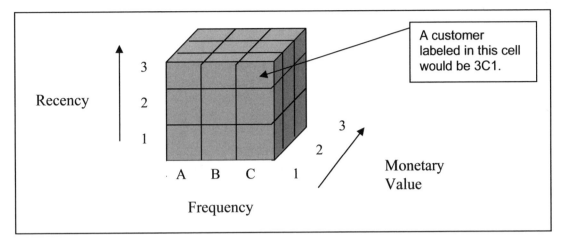

1.2.4 Purchase Affinity Clustering

A product manager may want to understand his customers based on their affinity for certain groups of products they have purchased within a certain time frame. To see this more clearly the manager computes an affinity score for the products of interest or perhaps all product categories, and then clusters those scores for similar groups. Another method of doing this is to cluster customers based on revenue and other demographics of interest and then score the product affinity for the cluster groups to observe whether there are any product tendencies for the customer segments. These kinds of clustering methods will be discussed in Chapters 9 and 10.

1.3 Typical Uses of Segmentation in Industry

In industry, segmentation or some sort of classification scheme has a wide variety of uses. A biologist might take field measurement samples and cluster them to find a useful taxonomy (Fisher 1936, pp. 179–188). In the medical field, clustering has been used to classify image data from Magnetic Resonance Imaging (MRI) scans for the purpose of detecting breast cancer (Getz, et al. 2003, p. 1079, 1089). In bioinformatics, a computer scientist working with a molecular biologist or a geneticist may seek to understand the function of genes. They may use genetic expression profile data and perform a hierarchical clustering in order to explore the structure of normal versus melanoma genes for the purpose of finding which genes may be responsible for the melanoma (Seo and Shneiderman 2002, pp. 80–86). In astronomy, measurements of star temperature and luminosity and X-ray or gamma ray emissions and other stellar sources are clustered to find similar star groups to aid in the understanding of the life cycle of stars; see Figures 1.3 and 1.4 (Berry and Linoff 1997, pp. 188–189).

When this clustering is performed on the star data, apparent distinct groups shown in Figure 1.3 appear that have specific attributes. White dwarf stars are late-stage stars that have shed off their outer layers. Red giants are stars that are middle- to late-stage stars that have swelled in size, and some can even migrate into supernova explosions. These clusters were profiled after much observation to ascertain these facts, and now a star classification system exists based on temperature and luminosity. A simplified cluster map of Figure 1.3 is shown in Figure 1.4 indicating three major star clusters.

Figure 1.3 Hertzsprung-Russel Diagram: Star Clusters by Temperature and Luminosity

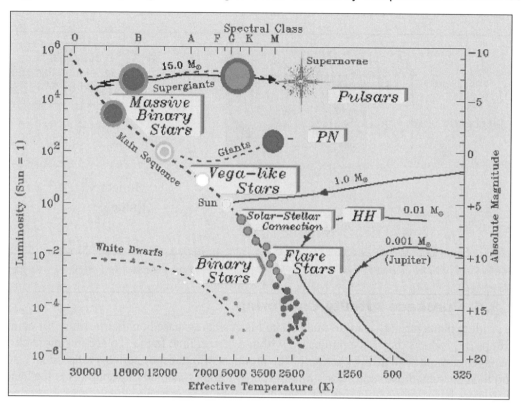

Figure 1.4 A Simplified Hertzsprung-Russel Diagram

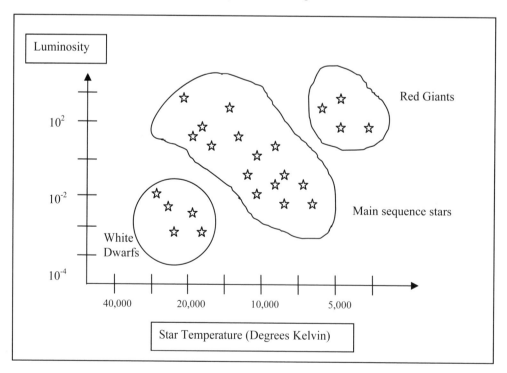

In marketing, an analyst may desire to classify customers according to similar customer groups for the purpose of understanding how to market to each customer segment. An analyst may want to classify research findings gathered from Web sites and other electronic means. To do so will cluster documents into themes without the analyst having to read each document and manually classify the documents into an organizational taxonomy. We will review this segmentation technique in Chapter 12, "Segmentation of Textual Data." In manufacturing, an engineer may want to better understand the mechanism or the root origin of a defect, so to aid this understanding the engineer clusters and sorts the defected items into similar defective categories. Cluster segmentation can be used to associate factor X with factor A, and a series of interconnected ideas may suggest models for the underlying mechanisms generating the observed data. In other words, cluster analysis may be used to reveal the structure and relations contained in the data (Anderberg 1973, p. 4). As you can see, there are many uses in industry where one can perform a classification according to predefined rules or a set of attributes, or segmentation of data into similar groups.

1.4 Segmentation as a CRM Tool

Segmentation is a set of techniques that can be beneficial in classifying customer groups. Typical direct marketing activities seek to improve the relationships with current customers. The better you know about your customer's needs, desires, and their purchasing behaviors, the better you can construct marketing programs designed to fit those needs, desires, and behaviors. Let's consider an example of a country variety store. In a southern New Hampshire town where I live, we have a country variety store that is an independent family business (not part of a franchise or chain). This store has a small delicatessen that offers sausage or meatball subs among other things. One of the unique aspects of these subs is the tomato sauce which is homemade. They offered these subs only on Wednesdays. In my opinion, they are one of the best sausage or meatball subs I've ever enjoyed. Therefore, when my family or I want a sausage or meatball sub, we would choose this small deli over any franchise stores available in town. The demand for these homemade subs caused the deli to offer their famous subs on each day rather than just one day of the week. How did the owners determine to move from offering these great subs on just Wednesdays to all days of the week? The answer is very straightforward in that they *observed* the demand of the subs and the requests made from various customers to offer these special subs on other days of the week. The owners, in fact, performed a *mental* segmentation as opposed to one with customer data in a database to reflect two apparent facts: 1) that the demand of these subs was higher than other products offered and 2) that the customers requested this service. So, the two facts put together made up the business decision to offer the subs more days of the week and thus better fulfill their customers' needs and desires; the simple supply-and-demand business curve. This simple example is what most direct marketers would like to achieve as well; however, one cannot segment a set of customers in a large database mentally as this country store owner did. However, with data mining algorithms such as clustering, decision trees, and other analytic tools, even when a business contains millions of customers the capability exists to group and segment these customers so that the segments are distinct groups of customers.

In the catalog industry, this kind of segmentation can be rather demanding. Take, for example, a large catalog mail-order company that has approximately 19 million customers. Their product offering is so large that they cannot offer all of their product offerings to all 19 million customers, especially in a single catalog. To do so would be cost prohibitive and the customer would have to search a huge catalog to find the items they desire. Therefore, the cataloger takes all of their customer data, attributes of these customers, and clusters them into differing segments containing various numbers of customers in each segment. Then, after profiling each of these customer segments, they offer a catalog designed specifically for each segment. A catalog for a teen segment would be very different from the one designed for middle-aged adults. This is not quite a one-to-one customer touch approach but a one-to-many approach, which is manageable and increases their customers' responsiveness to the catalogs offered in each segment (SAS Institute Inc. 2000, pp. 22–23).

In another example, a retail bank desires to improve their revenues and thus their profitability by segmenting their customer data according to the portfolio of products and services they have purchased. By clustering the customer data certain distinct patterns in one of the clusters appear—middle-aged customers who have a checking and savings account with fairly healthy balances, young customers who take advantage of more recent technological innovations, and older customers who could use some retirement plans, etc. This type of analysis and the set of business marketing ideas when brought together can make up the direct marketing activities and programs to leverage the cross-selling and up-selling of the bank's customer base and thus improve the revenue stream and also address customer loyalty.

Holding on to good customers and building up lesser customers is a common technique in direct marketing to generate more revenues and increase the breadth and depth of the products and services your customers will purchase. If you are a credit card company, then card profitability is achieved by balancing revenue (or reduce costs) against the company's risk. One method of revenue and risk segmentation splits revenue and risk and then profiles customers within these splits to observe any outstanding differences in the profile attributes. The data set from the northwest customer example in Tables 1.2 and 1.3 can be split into a simple segmentation of risk index and revenue classification. The risk index is a code from 00 to 05 or a null value. The code of 00 means no risk, 01 is relatively no risk, 02 is average risk, 03 is moderate risk, 04 is high risk, and 05 is very high risk. The revenue was sectored into low, medium, and high values. The following code produces the output in Figure 1.5.

Code Used to Generate Output in Figure 1.5

```
data work.nw_sales;
 length rev_class $12;
   set chapt1.northwest;
     if sales <= 10 then rev_class='Low Revenue';
     if sales >10 and sales < 5e4 then rev_class='Med Revenue';
   if sales >= 5e4 then rev_class='High Revenue';
run;

title 'Simple Segmentation of Risk Index by Revenue Class';
title2 'Northwest Customers Example Data Set';
proc freq data=work.nw_sales;
   table risk_index_code * rev_class /nocol norow nopercent nocum;
run;
title;
title2;
```

Figure 1.5 Simple Segmentation of Risk Index by Revenue Class

	Table of Risk_index_code by rev_class			
Simple Segmentation of Risk Index by Revenue Class **Northwest Customers Example Data Set** *The FREQ Procedure*				
Frequency				
Risk_index_code	**rev_class**			**Total**
	High Revenue	**Low Revenue**	**Med Revenue**	
00	122	110	692	924
01	14	75	13	102
02	3	20	5	28
03	0	4	2	6
04	0	3	2	5
05	0	1	0	1
Total	139	213	714	1066
Frequency Missing = 166				

As one might expect, the higher risk scores are mostly with low and medium revenues and little risk for high revenue customers. Perhaps in marketing to customers of low and medium revenue with high risk, an offer could be designed for them, and if leasing or credit is needed, a higher credit rate would be required for these customers than for customers with much lower risk. This is a simple segmentation using only two attributes, revenue and risk. We'll discuss type of segmentation in greater detail in Chapter 4.

Common sense would tell us that one of the first steps in successful CRM is to understand your customer. Just like the example with the country deli, the owners understood their customers' needs, desires, and spending habits. This information in turn led the owners to change their product offerings and frequency to better satisfy the customer. This simple fact of common sense does not always exist in many corporations. Many companies still do not see the value of their customers and the fact that their corporation exists *because* of their customers. The ones that do see this are hopefully trying to understand their customers and thus the techniques described in this book should aid the data miner, business analyst, marketer, etc. to know how to approach segmenting their customer base so that effective marketing can be administered to create an improved revenue stream and greater customer retention. In Chapter 2, a review of the underlying motivations for segmentation and descriptive-based models for your customers or prospects will be presented.

1.5 References

Anderberg, Michael R. 1973. *Cluster Analysis for Applications*. New York and London: Academic Press.

Berry, Michael J. A., and Gordon S. Linoff. 1997. *Data Mining Techniques: for Marketing, Sales, and Customer Support*. New York: John Wiley & Sons.

Fisher, Ronald Aylmer. 1936. "The Use of Multiple Measurements in Taxonomic Problems." *Annals of Eugenics* 7:179–188.

Getz, Gad, Hilah Gal, Itai Kela, Daniel A. Notterman, and Eytan Domany. 2003. "Coupled Two-Way Clustering Analysis of Breast Cancer and Colon Cancer Gene Expression Data." *Bioinformatics* 19.9:1079, 1089.

Hand, David J., Heikki Mannila, and Padhraic Smyth. 2001. *Principles of Data Mining*. Cambridge, MA: MIT Press.

SAS Enterprise Miner, Release 5.2. Cary, NC: SAS Institute Inc.

SAS Institute Inc. 2000. "Segmenting Customer Needs with Enterprise Miner." *SAS Communications* Q3: 22–23.

Seo, Jinwook, and Ben Shneiderman. 2002. "Interactively Exploring Hierarchical Clustering Results." *IEEE Computer, Special Issue on Bioinformatics* 35.7:80–86.

Why Segment?
The Motivation for Segment-
Based Descriptive Models

2.1 Mass Customization Instead of Mass Marketing

Why try and segment customers and attempt to treat various groups of customers differently than other groups? Why not treat all customers the same? In the 1970s and even into the early 1990s, marketers did much mass marketing. I can remember when as a youngster, the Sears catalog arrived at our home. It was the size of two volumes of the Encyclopaedia Britannica in thickness and had everything imaginable that Sears could offer its customers. The cost of mailing was much less in those days, but even with a large catalog, the cost was probably substantial. Eventually, the cost of mailing an entire catalog to many households across North America became too expensive and the profits dwindled. The example of the large Asian cataloger in Chapter 1, "Introduction," shows that one can obtain a much better return on an investment by tailoring the catalog to the sector of customers most likely to purchase the designed selection of goods. What marketers now need is mass customization instead of mass marketing. Customization means that for each individual customer, a product, promotion, offering, or service is tailored to that customer's needs or desires. This type of marketing is often referred to as one-to-one marketing (Peppers and Rogers 1997; Pine and Gilmore 1999). Many times, it is not possible to market exactly on a one-on-one basis. It is just too costly to design and mail a customized catalog for each individual. If printing could be customized as such in a very low-cost scenario, then this type of one-to-one marketing could exist. However, as in a catalog company, sometimes a one-to-few or one-to-segment is often a fair compromise to the mass marketing where everyone gets exactly the same offer or promotion. There are now some print shops that do customized printing at a low cost and so this one-to-one capability may be more realized than ever before. This is why e-mail marketing has become popular, because the cost associated with designing separate offers or communications is inexpensive and the delivery cost is also inexpensive as compared to more traditional media methods such as direct mail or telemarketing. The Internet is also a medium in which mass customization can take place at a relatively low cost. The Internet has tremendous growth not only in the number of sites available but also for the ability to do more accurate Web searches with search engines such as Yahoo, Google, AltaVista, and others. InternetWorldStats.com is a Web site, among many other sites, that posts statistics of Web usage around the world. The following table, as taken from this Web site in August 2006, shows typical usage statistics. A good portion of this Internet usage is being used for marketing of products and services both in the business-to-consumer and in business-to-business marketplace. Notice the usage growth from 2000 to 2005!

Table 2.1 World Internet Usage and Population Statistics

World Regions	Population (2006 Est.)	Population % of World	Internet Usage, Latest Data	% Population (Penetration)	Usage % of World	Usage Growth 2000-2005
Africa	915,210,928	14.1 %	23,649,000	2.6 %	2.3 %	423.9 %
Asia	3,667,774,066	56.4 %	380,400,713	10.4 %	36.5 %	232.8 %
Europe	807,289,020	12.4 %	294,101,844	36.4 %	28.2 %	179.8 %
Middle East	190,084,161	2.9 %	18,203,500	9.6 %	1.7 %	454.2 %
North America	331,473,276	5.1 %	227,470,713	68.6 %	21.8 %	110.4 %
Latin America/Caribbean	553,908,632	8.5 %	79,962,809	14.7 %	7.8 %	350.5 %
Oceania/Australia	33,956,977	0.5 %	17,872,707	52.6 %	1.7 %	134.6 %
WORLD TOTAL	6,499,697,060	100.0 %	1,043,104,886	16.0 %	100.0 %	189.0 %

NOTES: (1) Internet Usage and World population Statistics were updated for June 30, 2006. (2) Click each world region for detailed regional information. (3) Demographic (population) numbers are based on data contained in the world-gazetteer Web site. (4) Internet usage information comes from data published by Nielsen//NetRatings, by the International Telecommunications Union, by local NICs, and other reliable sources. (5) For definitions, disclaimer, and navigation help, see the Site Surfing Guide. (6) Information from this site may be cited, giving due credit and establishing an active link back to www.internetworldstats.com. ©Copyright 2006, Miniwatts Marketing Group. All rights reserved.

Let's take a look at a typical mass marketing technique. A marketing company surveyed how some companies spend their marketing money and found that 29% (the lion's share) of all marketing services are classified as a *mass marketing* segment. According to Levey (2002), these groups of mass marketers are very different from the other segments. The products and services sold by these marketers are typically characterized as very low cost and have a high turnover rate. Furthermore, these marketers generally assume that consumers do not want to form business relationships with companies that supply items such as laundry detergent, cat food, and instant breakfast drinks. Again, the cost of performing targeted marketing for these kinds of goods and services is too high for their return on investment. However, Levey points out a few very important items that could change this view dramatically. The growth of cable, satellite, and high-speed Internet are beginning to produce a more complicated marketing framework and choices for the mass marketers. This development has caused them to re-orient their thinking. Here are a couple of cases that point out that this targeted marketing can be very profitable and the concept of mass marketing is dwindling in favor of mass customization:

"Kraft General Foods has been one of the pioneers in this area. With a huge 30-million-name database begun with the Crystal Lite Lightstyle Club, Kraft targets diet-conscious consumers interested in fitness. Using newsletters, discount coupons and a catalog, they promote not only food products, but watches, mugs, jogging suits and other gear bearing the club logo (Levey 2002)."

"Information Resources, Inc. surveyed 7,900 shoppers and discovered that packaged-food companies were overspending on their Web sites in providing features customers didn't want. What consumers wanted was the ability to rate products and get coupons. What they didn't want were games and chat rooms. Half of all shoppers want coupons and free samples, but only 22 percent of Web sites offered them. Although only 38 percent of Web sites ask for feedback, a surprising 74 percent of respondents said they would provide it online (Levey 2002)." What customers desire should be the motivation that drives marketing to offer choices to the customers in a way that informs them as well as directs them to products and services.

For many marketers the ultimate goal is to market to a segment of one. The Internet and e-mail as communication platforms allow mass customization because these mediums have the potential for addressing and personalizing each customer individually. However, this does not mean that all Internet or e-mail marketing campaigns can create customized content for each individual customer. Consider again the case in Chapter 1 where the delicatessen in New Hampshire offered sausage or meatball subs with homemade spaghetti sauce to a group of customers who really wanted the kind of customized sub sandwich that has good homemade taste. However, mass customization is more of a delivery mechanism than it is a marketing concept. One-to-one marketing is clearly based on the idea of interacting with individual customers. Market segmentation, on the other hand, involves product development, message delivery, and distribution to groups of customers. Marketing is concerned about how to deliver the right messages to customers and even if you could use e-mail or the Internet for delivery, you may not be able to make up a separate message for each individual customer in your database (Levey 2002). If you have 600,000 unique customers in your database, then you would need to create 600,000 customized messages for each one of them in order to perform exact one-to-one marketing. So what is a marketer to do? You can probably find a group or segment of customers who are very similar in their demographics, purchasing behaviors, and even their attitudes or desired set of particular choices.

Progressive Insurance began its Autograph program in 1998. Through market research, Progressive learned that a segment of its customers valued an insurance program based on their personal driving experience. The system uses cellular and Global Positioning System technology to track actual driving patterns. Bob McMillian of Progressive maintains, "Our premise is that how you actually use and operate your car is more relevant to insurance pricing than traditional factors such as your gender, age or marital status." Progressive has, in fact, mass customized their auto insurance product (Levey 2002).

These segment groups then make up the sectors that the marketing department can send customized messages to in order to market better to those needs, desires, etc. This leads us into promotions and communications for various segment groups.

2.2 Specialized Promotions or Communications by Segment Groups

If one-to-one marketing is not possible due to high cost or ability to customize for many different individuals, then perhaps a one-to-segment group might be possible. If you know that customers in a particular group or segment have the right characteristics or affinity for a certain product or service, then you would naturally offer those customers the product or service that fits their needs the most. The basic idea behind a segment-based promotion or communication is that the customers in that segment have something in common and that something is what you want to exploit for the purpose of marketing directly to them. The key to this kind of marketing program is to know and understand the common features of the customers (or perhaps prospects) in the segment of interest.

Let's say we have performed some sort of segmentation on business customers and we now wish to begin some marketing programs with the segments we have of our customer base data. The customer data has been segmented into three segment groups as given in Table 2.2. Your company is a consulting and customized teaching firm, and you want to best market your services to each of the three segments in Table 2.2. The services you provide are training classes, both in house or at training centers, consulting services, and training materials. The underlying question that the business and marketing team needs to address is: what types of marketing programs should we develop for each of these three segments? Look at the segment profile for each of the three segments in Table 2.2. Are there distinct differences in these segments that allow marketing to better develop a program that is tailored to the segments under consideration? What would be a good set of programs targeted and aimed expressly for each of these segment groups?

Table 2.2 Three Segments of Customers for Marketing

Segment Number	Customer Profile
1	This group of customers (2564 unique customers) is made up of mostly medium-sized companies (average corporate-level employee size is 450 employees). They purchase mostly through direct channels; however, about a third also purchase through a reseller. They are mainly made up of the financial services industry, including insurance. This group has been typically a loyal group of customers purchasing for an average of 5.4 consecutive years. This group uses your training services highly and also consulting services to aid their in-house training and project needs.
2	Segment of customers (12,344 unique customers) is made up of rather small-sized companies (average number of corporate employees is 50). They purchase only a few of your teaching products primarily through your resellers and distributors (about 65%). They don't purchase any of your consulting services, and these customers are relatively new, purchasing an average of 1.5 consecutive years. These customers are mainly made up of small personal services industries that are typically a strong growth sector. They primarily purchase only your training materials.

(continued)

Table 2.2 (*continued*)

Segment Number	Customer Profile
3	This group of customers (821 unique customers) is made up of very large companies (average number of corporate employees is 1100). They are not very loyal customers; however, they do purchase a fairly good breadth of your products and consulting services. They have purchased consecutively for an average of 2.5 years with a good number not purchasing anything in over one to two years. These customers are large chemical manufactures, and they have the potential to purchase more from your company as they also purchase some from your competitors as well. This group uses some consulting and training services, and they also do purchase training materials. One hundred percent of this segment purchases direct only.

With the profiles of the segments in Table 2.1 at hand, let us see if we can devise some specific marketing programs that would be tailored to each group. In segment 1, we find that the characteristics that typify this group are mostly direct purchases, rather loyal, financial service industry and insurance, and they are medium-sized companies. We also know that both training services and consulting services are used well. It might be rather straightforward to offer more consulting services for this group and for the third that purchases from your resellers and partners, and to add an offer for the resellers to use for their end-user customers to entice greater consulting offerings. Since these are mostly financial and insurance related companies, it may be a good idea to make the offer of consulting services specific for those industries. For segment 2, perhaps a special offer for consulting might be in order as they don't typically use that service and it would be good to increase loyalty as this is a young customer group with respect to segment 1. Segment 3 has not typically purchased anything in one to two years so offers to prevent customer attrition would be beneficial here. We will get into much more detail on how to estimate product affinities by segment later in Chapter 9, "Clustering and the Issue of Missing Data."

This segmentation is often referred to as behavioral segmentation as the demographics are derived from purchase transaction history and other general demographics. Attitudinal segmentation is one in which the characteristics are the desires or preferences that the customer has indicated through surveys, responses to marketing programs or offers, or information gathered through e-mail or call center representatives. The best of both worlds would be ideal, which would be to have both kinds of information in the database; however, this is often not the case. Behavioral data is typically more available than attitudinal data; however, it would be prudent to collect both types of data in your customer database. Having both attitudinal and behavioral data might allow you to create models of the attitudinal data with the customer demographics and behavioral patterns and to project those attitudes onto the remaining set of customers where no attitudinal data exists or perhaps on a prospect database.

2.3 Profiling of Customers and Prospects

One of the main issues in any segmentation is the profile of the segment under question. The profile of that segment is the basic description of the common elements that each customer or prospect shares within that segment. Comparing and contrasting segment profiles allows one to understand better the set of customers represented in each segment. So how does one go about profiling a set of customers? The answer starts in a data assay. The word *assay* in the *Oxford*

English Dictionary is "the trying in order to test the virtue, fitness, etc. (of a person or thing)." This is what we want to do with data so the data assay produces detailed knowledge, and is usually a report of the quality, problems, shortcomings, and suitability of the data for mining (Pyle 1999, p. 125). The aspects of a data assay typically start with some basic characteristics like the number of unique values for a categorical variable, the percentage of missing values, mean and standard deviation and outliers for numeric variables. This kind of summary, being tabulated in a kind of report, allows one to survey the variables quickly and can give the analyst clues about how to approach certain kinds of data mining. This report can form the foundation for all preparation and mining work that follows. For example, if two categorical variables or columns in a data set each have 10% missing values, then when combinations of these variables are used, the amount of missing of the combined data can be much more than 10% depending on the overlap of the two fields when combined. This type of difficulty has a large negative affect on the outcome of data mining or statistical analyses and one needs to be cognizant that this can happen more often than not on typical data sets. So, let's begin working with an example to demonstrate a preliminary data assay and start a profile exercise.

In all of the exercises in this book, I will outline a brief process flow table like the following one. The process flow table indicates the major steps that are to be taken to complete the data mining exercise.

Process Flow Table: Data Assay Project

Step	Process Step Description	Brief Rationale
1	Start SAS Enterprise Miner and create a new project.	
2	Add a data source to the data mining flow – BUYTEST data set. Use the data source wizard in SAS Enterprise Miner.	BUYTEST data set is explained in Appendix 1.
3	View data columns in the BUYTEST data set. Find where the number of rows/columns is listed in the data source property sheet.	First look at the data in the Data Assay process.
4	Review the distribution of the PURCHTOT column on BUYTEST.	Shows how variables are distributed.
5	View another column INCOME in the BUYTEST data set.	Shows how variables are distributed.
6	Change the role of the variable RESPOND to a Target attribute.	Modifies variable attributes in SAS Enterprise Miner.
7	Add a Stat/Explore and MultiPlot node to the data mining flow.	Performs basic statistics and plots of variables in the Data Assay process.
8	Observe the results from MultiPlot and Stat/Explore nodes.	Makes general statistical observations on variables of the data set.
9	Use Worth stats to understand the target RESPOND variable.	Shows how variables relate to the target variable.
10	Add a SAS Code node and generate simple one-way distribution tables.	Relates customer attributes from the basic frequency statistics – Data Assay profile.
11	Create crosstabulation distribution tables.	Relates customer attributes from the basic frequency statistics – Data Assay profile.

(continued)

Process Flow Table (*continued*)

Step	Process Step Description	Brief Rationale
12	Change the role of CLIMATE and DISCBUY variables to "Segment."	Allows the Segment Profile node to profile.
13	Change the "Use Segment Variable" property for more profiling in Stat/Explore node.	Additional techniques for Data Assay profiling – shows impact of variables on the target RESPOND variable.

2.3.1 Example 2.1: The Data Assay Project

To invoke SAS Enterprise Miner double-click the EM 5.2 Client icon on your desktop. If you cannot find the icon, then go to the Start menu and select **Smart→Programs→SAS, Enterprise Miner→EM 5.2 Client**. A window will appear that asks you to select the Personal Workstation. This window looks like the one depicted in Figure 2.1. Select the Personal Workstation if that is your configuration setup. After you log on with your user name and password, the next screen (shown in Figure 2.2) shows the SAS Enterprise Miner startup window. This is where you can open an existing project or start a new one. Copy the BUYTEST data set in the Chapter 2 folder to the SAMPSIO SAS data library location. We will use a node within SAS Enterprise Miner to demonstrate the data assay; however, much of these could also be generated using Base SAS procedures and DATA step statements.

Figure 2.1 Starting SAS Enterprise Miner

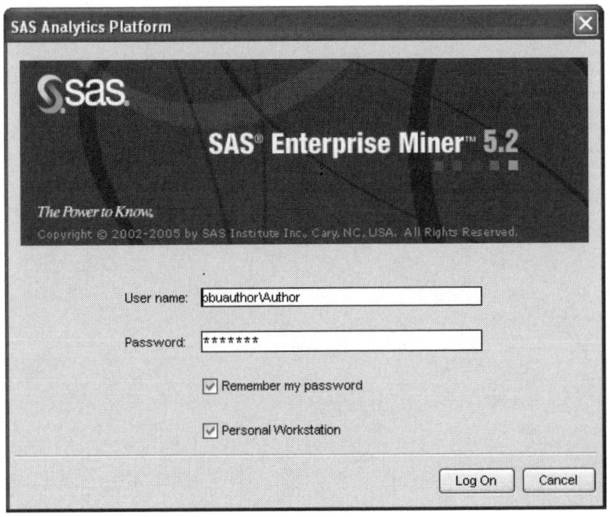

Figure 2.2 Initial SAS Enterprise Miner Startup Window

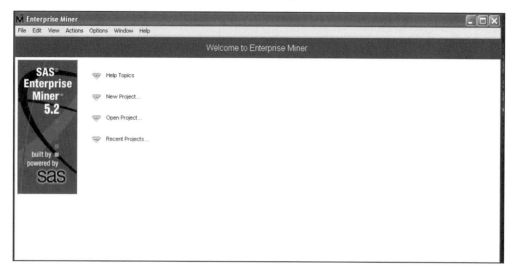

Step 1: Create a new project and call it "Data Assay Profile." Your window, at this point, should look like Figure 2.3.

Figure 2.3 Beginning Data Assay Project

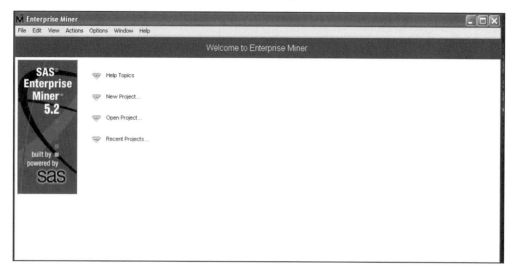

Double-click the Data Sources icon, and follow the instructions to add the BUYTEST data from the SAMPSIO library. When you get to the section that asks you to select the Apply Advisor Options (Basic or Advanced), select **Advanced**. Continue to select **Next**, and use the default settings for variables settings. Set the role of the data set to **Raw**. In the next example we'll create a "Profile" process flow as shown just below the "Data Assay" flow diagram. Now create a new flow diagram and call it "Data Assay." **Step 2:** Drag the BUYTEST data set from the Data Sources icon onto the diagram workspace. Your diagram should now look like Figure 2.4.

Figure 2.4 Example 2.1 Data Assay—BUYTEST Input Data Source

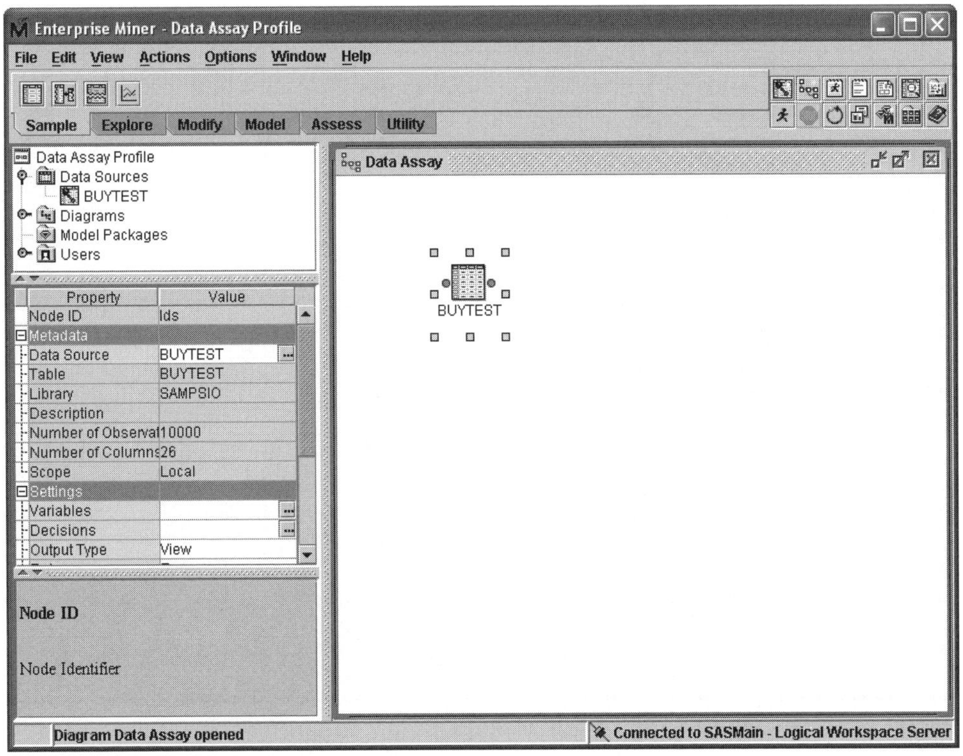

Step 3: Now click **Variables** in the Property-Value section and your window should now look like that in Figure 2.5. Notice that the number of columns (variables) and rows (observations) in your data set are listed. In many other data sets where you have many fields, a scroll bar appears on the right side so you can view all the fields. This view can give you a brief understanding of the types of variables you have in your data set, if they are nominal or interval, and the measurement type as well. SAS Enterprise Miner has attempted to make a best guess at the Level column, which indicates how SAS Enterprise Miner treats the variable. For example, notice the variable CLIMATE is considered nominal. This means that the values seen by the Data Advisor are a categorically grouped set of values. If you believe that the 10–30 values should be ordinal, you could change that at this point and SAS Enterprise Miner would treat this variable throughout the process flow as ordinal unless you decide to modify its attributes. To perform one of the first set of data assay reports, SAS Enterprise Miner provides you with a few methods that aid in the data assay process. Select the CLIMATE variable and then click the **Explore** button. This view gives you some very basic descriptive statistics about the CLIMATE variable in your data set. Note, however, that these statistics are based on the default samples when the BUYTEST data source was added by the Data Advisor. The default number of rows that are used is 2,000 rows, which are sampled at random. A simple histogram is given and you have further options of plotting more data to start your data assay exploratory analysis.

Figure 2.5 Data Assay—Variables in the BUYTEST Data Set Node

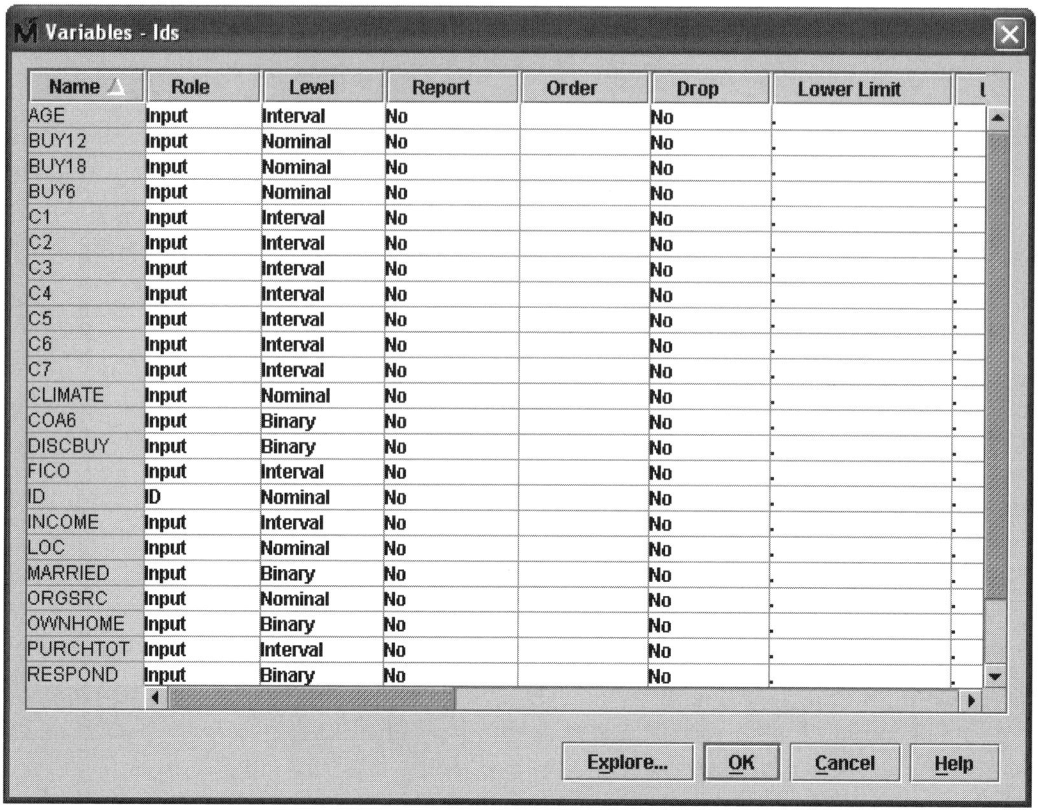
Variables - Ids
Name
Role
Level
Report
Order
Drop
Lower Limit
AGE Input Interval No No
BUY12 Input Nominal No No
BUY18 Input Nominal No No
BUY6 Input Nominal No No
C1 Input Interval No No
C2 Input Interval No No
C3 Input Interval No No
C4 Input Interval No No
C5 Input Interval No No
C6 Input Interval No No
C7 Input Interval No No
CLIMATE Input Nominal No No
COA6 Input Binary No No
DISCBUY Input Binary No No
FICO Input Interval No No
ID ID Nominal No No
INCOME Input Interval No No
LOC Input Nominal No No
MARRIED Input Binary No No
ORGSRC Input Nominal No No
OWNHOME Input Binary No No
PURCHTOT Input Interval No No
RESPOND Input Binary No No
Explore...
OK
Cancel
Help

Figure 2.6 shows the histogram of the CLIMATE variable.

Figure 2.6 Climate Sample Distribution from Selecting Explore Option

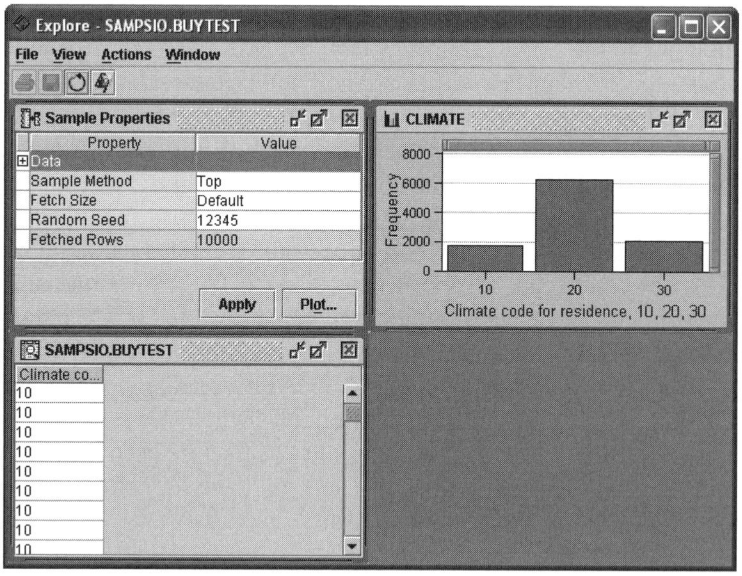
Explore - SAMPSIO.BUYTEST
File View Actions Window
Sample Properties
Property
Value
Data
Sample Method Top
Fetch Size Default
Random Seed 12345
Fetched Rows 10000
Apply
Plot...
CLIMATE
8000
6000
4000
2000
0
Frequency
10
20
30
Climate code for residence, 10, 20, 30
SAMPSIO.BUYTEST
Climate co...
10
10
10
10
10
10
10
10

If you have a rather large data set, you can always select the number of rows that may better represent the overall data or you can also select all the rows in which to perform the analysis, but this will take quite a bit longer than just doing the sample. Remember, the purpose of the data assay is to give you a first look at your data and to see what your data is made of and how appropriate it is for data mining. Sampling is typically good for this kind of first look analysis.

Step 4: Now let's consider the PURCHTOT variable. Highlight this variable and use the **Explore** button as we did before. Figure 2.7 shows the distribution of total purchases. Notice how skewed this distribution is (i.e., it is not *normal* or Gaussian in any way).

Figure 2.7 Sample Distribution of PURCHTOT in the BUYTEST Data Set

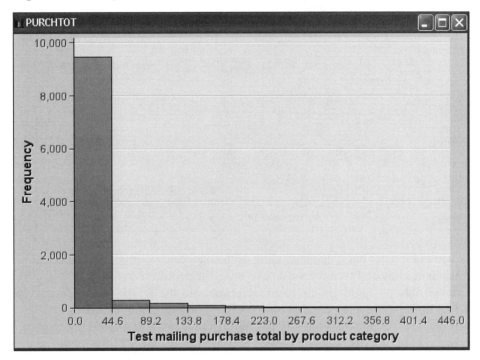

You can drag the corner of the plot frame to enlarge or reduce the size of the plot. From this plot in Figure 2.7, the PURCHTOT variable is highly skewed to the left. Since this data is highly *non-normally distributed* this can affect how certain data mining algorithms work on this variable. We will investigate this in more depth when we perform clustering algorithms later in Chapter 5, "Segmentation of Several Attributes with Clustering." The method for obtaining the data assay can be documented in the following fashion (not necessarily order specific). **Step 5:** Now explore the INCOME variable. This is shown in Figure 2.8.

- Survey the number of rows in your data set and the number of variables (or fields). This should match what you believe your data set should contain.

- Review the means and standard deviations of the interval data as this give you an idea of the placement of each variable or field in the data space.

- Review the number of missing data elements for each variable. If you use combinations of variables with missing data, then perhaps data imputation might be needed.

- Review the distributions of several variables to observe how these variables are distributed. For example, in Figure 2.9 the INCOME variable appears to have a non-normal shape, especially on the left side of the distribution; however, the right side tapers off to a more normal shape. This may give indications of multiple distributions within the INCOME variable.

Figure 2.8 INCOME Distribution in the BUYTEST Data Set

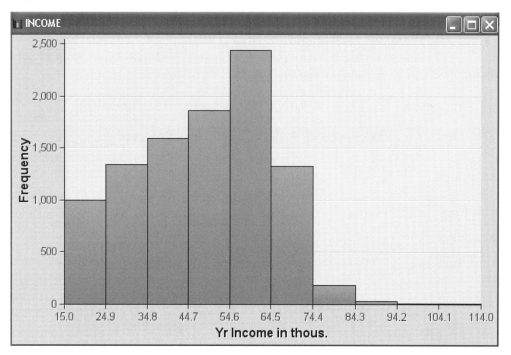

These steps will aid in your basic understanding of the data prior to mining. Another brief example in the data assay and then we will move on to the profile stage. **Step 6:** In the Variables of the BUYTEST data set, highlight the RESPOND variable and set the role to Target. This will indicate to SAS Enterprise Miner that this binary variable is our target response variable. **Step 7:** On the **Tools** tab, find the node labeled **Distribution Explorer**. Drag a StatExplore node and a MultiPlot node to your process flow work space and connect the BUYTEST Input Data source node to each of these nodes so that your diagram looks like the one in Figures 2.9 and 2.10. While highlighting the StatExplore node, set the Interval variable to **Yes** in the property sheet. Now connect a Control Point node and connect both the StatExplore and MultiPlot node to it so that your diagram looks like Figure 2.10.

Figure 2.9 StatExplore and MultiPlot Nodes

Figure 2.10 Use of Control Point Node

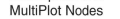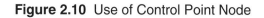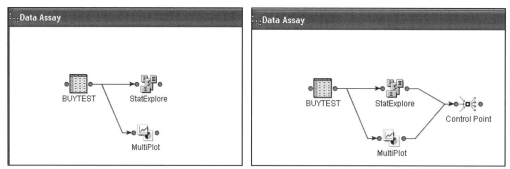

Now you can run both paths by right-clicking the Control Point node and selecting **Run**.
Step 8: Since we selected a target variable (and it happens to be binary), when you highlight the MultiPlot node and select results, you should see a bar chart plot of AGE versus the target variable RESPOND as shown in Figure 2.11.

Figure 2.11 MultiPlot Node Results of AGE vs. RESPOND Variable

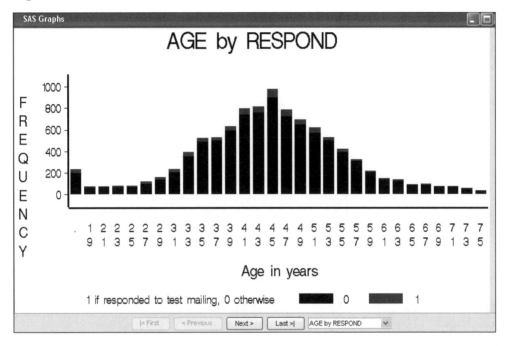

The plot in Figure 2.12 shows the combined effect of AGE and RESPOND, which gives a bivariate plot distribution of these variables. This becomes important when data mining models use the interaction of AGE and RESPOND as a variable in the model. You can select other variables to plot by choosing the bottom menu selection in this plot. These types of options allow you to continue your understanding of the overall data assay process.

Now, open the results of the StatExplore node. Select the **View** menu, and choose **Summary Statistics of Interval** variables and then **Categorical** variables. These tables describe some of the basic statistics about the data preceding the StatExplore node, in this case the BUYTEST data set. Figures 2.12 and 2.13 show the categorical (class) and numeric variable statistics respectively. From these basic statistics we can derive the following set of information from the BUYTEST data set:

- The SEX variable has about 2% missing data as well as OWNHOME.

- In the numerical variables, only AGE, INCOME, and FICO contain missing values and are relatively low percentages based on sample data.

- From the mean and standard deviations of the numeric variables, all variables are reasonably well behaved (i.e., no extreme outliers that are extremely large or small possibly indicating incorrect data points, like the maximum value for age being 250 or the minimum value of age being 8).

- The ORGSRC (original customer source) variable has the largest amount of missing data, about 5% based on the sample.

Figure 2.12 Categorical Variable Stats for the BUYTEST Data Set

Target	TARGETVA...	Variable	Type	Numeric Val...	Formatted V...	Frequency	Percent	Role	PERCENT	PLOT
RESPOND	0	ORGSRC	C		.O	1853	20.06932	INPUT	0.1853	
RESPOND	1	ORGSRC	C		.O	164	21.38201	INPUT	0.0164	
RESPOND	0	ORGSRC	C		.P	1350	14.62147	INPUT	0.135	
RESPOND	1	ORGSRC	C		.P	124	16.16688	INPUT	0.0124	
RESPOND	0	ORGSRC	C		.R	1102	11.93545	INPUT	0.1102	
RESPOND	1	ORGSRC	C		.R	90	11.73403	INPUT	0.009	
RESPOND	0	ORGSRC	C		.U	1542	16.70096	INPUT	0.1542	
RESPOND	1	ORGSRC	C		.U	134	17.47066	INPUT	0.0134	
RESPOND	0	OWNHOME	N		.	198	2.144482	INPUT	0.0198	
RESPOND	1	OWNHOME	N		.	36	4.693611	INPUT	0.0036	
RESPOND	0	OWNHOME	N	00	00	5952	64.46442	INPUT	0.5952	
RESPOND	1	OWNHOME	N	00	00	551	71.83833	INPUT	0.0551	
RESPOND	0	OWNHOME	N	11	11	3083	33.3911	INPUT	0.3083	
RESPOND	1	OWNHOME	N	11	11	180	23.46806	INPUT	0.018	
RESPOND	0	RETURN24	N	00	00	8573	92.85173	INPUT	0.8573	
RESPOND	1	RETURN24	N	00	00	718	93.61147	INPUT	0.0718	
RESPOND	0	RETURN24	N	11	11	660	7.148273	INPUT	0.066	
RESPOND	1	RETURN24	N	11	11	49	6.388527	INPUT	0.0049	
RESPOND	0	SEX	C		.	198	2.144482	INPUT	0.0198	
RESPOND	1	SEX	C		.	36	4.693611	INPUT	0.0036	
RESPOND	0	SEX	C		.F	4139	44.82833	INPUT	0.4139	
RESPOND	1	SEX	C		.F	350	45.63233	INPUT	0.035	
RESPOND	0	SEX	C		.M	4896	53.02719	INPUT	0.4896	
RESPOND	1	SEX	C		.M	381	49.67405	INPUT	0.0381	

Figure 2.13 Numeric Variable Stats for the BUYTEST Data Set

TARGET	TARGETVA...	Variable	LABEL OF F...	MEAN	STDDEV	Non Missing	Missing	Min	Median	Max
RESPOND	0	C7		0	1.863437	9233	0	0	0	
RESPOND	1	C7		2.275098	6.367238	767	0	0	0	6
RESPOND	0	C1		0	1.488087	9233	0	0	0	
RESPOND	1	C1		2.024772	5.011544	767	0	0	0	4
RESPOND	0	C3		0	4.240303	9233	0	0	0	
RESPOND	1	C3		7.346806	13.59426	767	0	0	0	12
RESPOND	0	C4		0	4.54936	9233	0	0	0	
RESPOND	1	C4		7.005215	14.99321	767	0	0	0	12
RESPOND	0	C5		0	2.92789	9233	0	0	0	
RESPOND	1	C5		4.615385	9.602514	767	0	0	0	9
RESPOND	0	C6		0	15.49322	9233	0	0	0	
RESPOND	1	C6		45.52282	34.89255	767	0	0	37	24
RESPOND	0	C2		0	6.28347	9233	0	0	0	
RESPOND	1	C2		13.19687	18.82471	767	0	0	0	11
RESPOND	0	PURCHTOT	Test mailing p...	0	27.27633	9233	0	0	0	
RESPOND	1	PURCHTOT	Test mailing p...	81.98696	59.13881	767	0	1	69	44
RESPOND	0	VALUE24	Total value of...	251.2976	153.4221	9233	0	60	213	125
RESPOND	1	VALUE24	Total value of...	287.1186	180.683	767	0	61	231	101
RESPOND	0	AGE	Age in years	44.76292	10.1065	9035	234	18	44	7
RESPOND	1	AGE	Age in years	42.01642	10.27857	731	36	18	43	7
RESPOND	0	INCOME	Yr Income in ...	48.01118	16.00472	9035	234	15	50	11
RESPOND	1	INCOME	Yr Income in ...	47.22298	16.64491	731	36	15	49	9
RESPOND	0	FICO	Credit Score	694.6629	28.79151	9196	39	577	696	80
RESPOND	1	FICO	Credit Score	690.2706	29.10468	765	2	604	691	77

Step 9: In the StatExplore node, there is another useful chart and that is the Worth statistic. This shows the relative importance of a variable to the Target variable (in this case the RESPOND variable). To view this graph, select **View → Plots → Variable Worth**. Figure 2.14 shows the variable Worth statistic; you can place the mouse pointer over each variable to see the variable name/label and Worth statistic. The PURCHTOT variable has the largest positive relative impact on the RESPOND variable.

Figure 2.14 Relative Variable Importance to the RESPOND Target

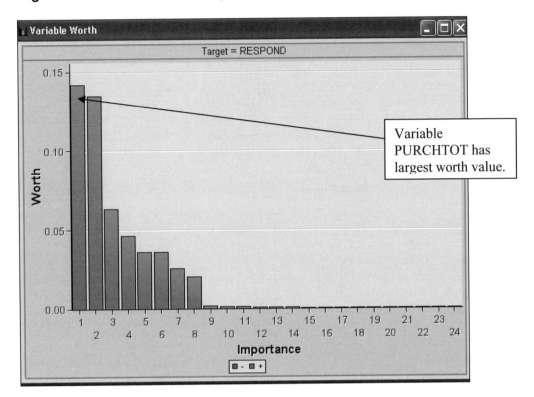

2.3.2 Example 2.2: Customer Profiling of the BUYTEST Data Set

A profile of customers may include or exclude various types of reports, graphs, plots, distributions, etc., depending on the type of data mining one needs to perform for the business problem or issue to be solved. One common item to start with is a simple one-way frequency distribution. This type of distribution can be easily generated by a SAS/STAT procedure called the FREQ procedure or from the output we generated in the MultiPlot node. **Step 10:** Using PROC FREQ, let's try it out on the BUYTEST data set. Add a SAS Code node after the Control Point and place the code as shown in the following example. The distributions of four variables: CLIMATE, BUY6, BUY12, and BWY18 are used in the TABLE statement of PROC FREQ.

Code Used to Generate Output in Figure 2.15

```
title 'Distribution of Climate Codes & Purchase Recency';
  proc freq data=emws.ids_data;
    table climate buy6 buy12 buy18 ;
  run;
title;
```

Figure 2.15 One-Way Distribution of Four Variables in the BUYTEST Data Set

```
Distribution of Climate Codes & Purchase Recency

The FREQ Procedure

            Climate code for residence, 10, 20, 30

                                   Cumulative   Cumulative
CLIMATE    Frequency    Percent    Frequency     Percent
----------------------------------------------------------
10            1700       17.00        1700        17.00
20            6257       62.57        7957        79.57
30            2043       20.43       10000       100.00


                    # of purchases 6mo

                                   Cumulative   Cumulative
BUY6       Frequency    Percent    Frequency     Percent
----------------------------------------------------------
0             8757       87.57        8757        87.57
1             1203       12.03        9960        99.60
2               40        0.40       10000       100.00


                    # of purchases 12mo

                                   Cumulative   Cumulative
BUY12      Frequency    Percent    Frequency     Percent
----------------------------------------------------------
0             8024       80.24        8024        80.24
1             1854       18.54        9878        98.78
2              121        1.21        9999        99.99
3                1        0.01       10000       100.00


                    # of purchases 18mo

                                   Cumulative   Cumulative
BUY18      Frequency    Percent    Frequency     Percent
----------------------------------------------------------
0             7001       70.01        7001        70.01
1             2550       25.50        9551        95.51
2              426        4.26        9977        99.77
3               23        0.23       10000       100.00
```

The SAS FREQUENCY procedure displays the distribution of number of unique items for each level of the variables requested, the percent, cumulative counts, and the cumulative percent. This type of profiling allows you to see that for 6–18 months, most customers (70–90%) have not purchased at all. Notice that almost two-thirds of the customers live in the climate categorized as "20," and 80% of the customer base did not purchase anything in the past 12 months. Using combinations of these variables, one can build crosstabulations as well. **Step 11**: Add the following additional code to the SAS Code node, and this SAS FREQUENCY procedure will generate the output as shown output in Figure 2.16 once you run the node.

Code Used to Generate Output in Figure 2.16

```
title 'Cross-Tab of Married and Residence Location';
  proc freq data=emdata.buytest;
    table loc*married ;
  run;
title;
```

Figure 2.16 Crosstabulation Results of Residence by Married

```
Cross-Tab of Married and Residence Location

The FREQ Procedure

Table of LOC by MARRIED

LOC(Location of residence, A-H)
        MARRIED(1 if Married, 0 otherwise)
Frequency|
Percent  |
Row Pct  |
Col Pct  |      0|      1|  Total
---------+--------+--------+
A        |    225 |    340 |    565
         |   2.30 |   3.48 |   5.79
         |  39.82 |  60.18 |
         |   5.54 |   5.96 |
---------+--------+--------+
B        |    735 |   1063 |   1798
         |   7.53 |  10.88 |  18.41
         |  40.88 |  59.12 |
         |  18.10 |  18.63 |
---------+--------+--------+
C        |    202 |    345 |    547
         |   2.07 |   3.53 |   5.60
         |  36.93 |  63.07 |
         |   4.97 |   6.05 |
---------+--------+--------+
D        |    226 |    319 |    545
         |   2.31 |   3.27 |   5.58
         |  41.47 |  58.53 |
         |   5.57 |   5.59 |
---------+--------+--------+
E        |    945 |   1253 |   2198
         |   9.68 |  12.83 |  22.51
         |  42.99 |  57.01 |
         |  23.27 |  21.96 |
---------+--------+--------+
F        |    933 |   1189 |   2122
         |   9.55 |  12.17 |  21.73
         |  43.97 |  56.03 |
         |  22.97 |  20.84 |
---------+--------+--------+
G        |    352 |    578 |    930
         |   3.60 |   5.92 |   9.52
         |  37.85 |  62.15 |
         |   8.67 |  10.13 |
---------+--------+--------+
H        |    443 |    618 |   1061
         |   4.54 |   6.33 |  10.86
         |  41.75 |  58.25 |
         |  10.91 |  10.83 |
---------+--------+--------+
Total        4061     5705    9766
            41.58    58.42  100.00

Frequency Missing = 234
```

The crosstabulation table in Figure 2.17 indicates that most of the married customers are in locations B, E, and, F, and make up about 60% of the customer base. Note that the combined set of Married vs. Location produces 234 missing data elements. These types of tabular reports aid in the understanding of the customer data. A similar type of report is available in the MultiPlot node. Open the results of the MultiPlot node and review the contents in the Output window. Although the output is not as easy to interpret as the FREQUENCY procedure displays, it does provide similar information.

Using these types of tools, one can obtain the data assay and profile information that begins to build an understanding of what the data is made of.

Another area for exploration involves profiling. SAS Enterprise Miner 5.2 contains a Segment Profile node. To use this node, however, a variable's role must be *cluster* or *segment*. So, highlight the BUYTEST input data set and click **Variables**. **Step 12**: Change the role of the CLIMATE and DISCBUY variables to **Segment**. Drag a Segment Profile node to your diagram and connect the BUYTEST to this node. Run the Segment Profile node with all default settings. In the Results window, you should see a similar output to that shown in Figure 2.18 highlighting the Profile window in the results.

Figure 2.17 Segment Profile Node Results

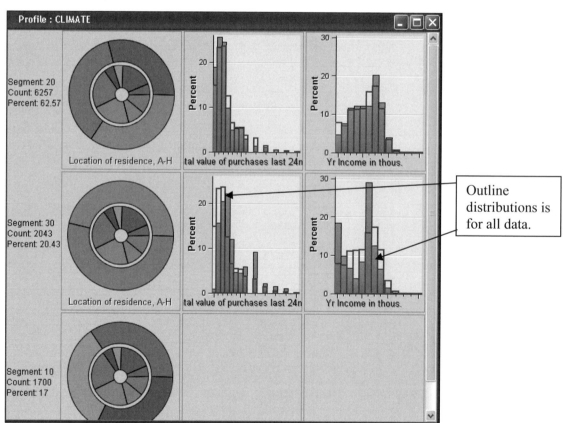

The output in Figure 2.17 shows the effect the CLIMATE variable has on the VALUE24 and INCOME distributions. The outline distribution curve is for all the data where the default blue distribution is for the segment only. The Segment Profile node also provides several other types of graphical reports. In the Results window, there should be a Variable Worth output. Figure 2.18 shows Variable Worth for the CLIMATE variable. This plot shows clearly that the LOC variable has much more influence on the target variable we selected (RESPOND) than do VALUE24 and INCOME variables.

Figure 2.18 Variable Worth Plot of CLIMATE in Segment Profile Results Window

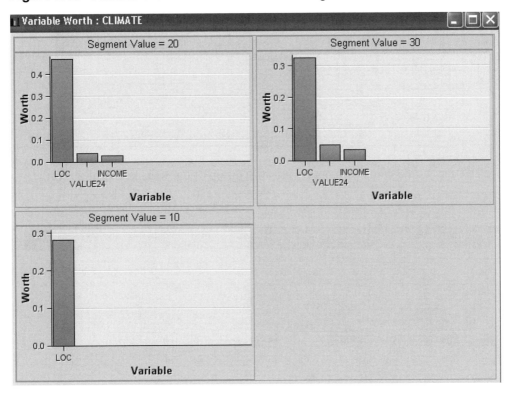

Another method of reviewing the variables for profiling is to go back to the StatExplore node. **Step 13**: Since we changed the variable roles on the BUYTEST data to *Segment* for CLIMATE and DISCBUY, change the **Use Segment Variable** property to **Yes** and rerun the StatExplore node. Open the StatExplore node Results window. You should now see a chi-square plot of the segment variables to the target variable. This is shown in Figure 2.19. Placing your mouse over each cell of the matrix plot allows you to observe the contribution of the category variable on the vertical axis and the numeric variable on the horizontal axis to the target variable RESPOND. For example, if you place your mouse over the intersection of PURCHTOT and CLIMATE:30 cell, the cell chi-square (labeled as Cramer's V) should be 0.6789. The darker the shaded cell, the larger the cell chi-square value and therefore the larger the contribution to the target variable RESPOND.

As can be seen from this set of respective analyses, a better understanding of what is in the data can be obtained from various combinations of profile and data assay results. There is much more that can be done in the preparation of the data and an excellent reference for this is *Data Preparation for Data Mining,* by Dorian Pyle (1999).

Figure 2.19 Chi-Square Plot of Segment Variables with Respect to Target Variable

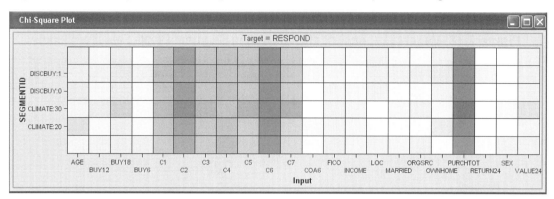

Figure 2.20 shows the completed process flow diagram of our data assay and profiling.

Figure 2.20 Completed Process Diagram of Data Assay and Profiling

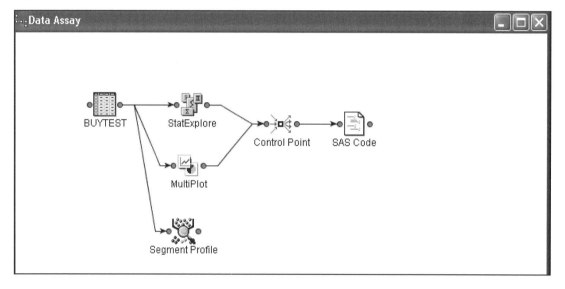

2.3.3 Additional Exercise

In the profiling exercise, change other variable roles in BUYTEST that are considered as nominal to a role of *Segment* instead of *Input*. Rerun the StatExplore and the Segment Profile nodes in the process flow diagram. Comment on the results with respect to the target variable RESPOND.

2.4 References

Enterprise Miner, Release 5.2. SAS Institute Inc., Cary, NC.

Levey, Doran J. 2002. "Segmentation: The Mass Market is Changing." *DM Review.* June.

Peppers, Don, and Martha Rogers. 1997. *Enterprise One To One: Tools for Competing in the Interactive Age.* New York: Currency Doubleday.

Pine, B. Joseph, and James H. Gilmore. 1999. *The Experience Economy: Work Is Theatre & Every Business a Stage.* Boston, MA: Harvard Business School Press.

Pyle, Dorian. 1999. *Data Preparation for Data Mining.* San Francisco: Morgan Kaufmann Publishers, Inc.

Distance: The Basic Measures of Similarity and Association

3.1 What Is Similar and What Is Not

Sometimes the phrase "look at the big picture" is used to back away from the details and look at what the *main* patterns or effects are telling you about the set of customers, patients, prospects, cancer treatment records, or measures of star temperature and luminosity. When a database may contain so many variables and rows or records and so many possibilities of dimensions, the structure could be so complex that even the best set of *directed* data mining techniques are unable to ascertain meaningful patterns from it. I use the term *directed* as most data mining techniques are classified into either *directed* or *undirected*. In directed data mining, the goal is to explain the value of some particular field or variable (income, response, age, etc.) in terms of all of the other variables available. A target variable is chosen to tell the computer algorithm how to estimate, classify, or predict its value. Undirected data mining does not have a predicted variable; instead, we are asking the computer algorithm to find a set of patterns in the data that may be significant and perhaps useful for the purpose at hand (Berry and Linoff 1997, pp. 188–189). Another way of thinking about undirected data mining or knowledge discovery is to *recognize* relationships in the data and directed knowledge discovery to *explain* those relationships once they have been detected. In order for an algorithm to find relationships, a set of rules or criteria must be made to measure the associations among the individuals so that patterns can be detected.

One way to think of similarity (or the converse, dissimilarity) is to devise a measure or metric, measure the items of interest, and then classify the items according to their measures. For example, a set of students in a class could be measured for their height. Once each student is measured, one could classify the students as tall, medium, or short. A secondary characteristic, sex, could then be added to the height measurement and now the measure of similarity is both height and sex in combination. This concept is shown in Figure 3.1.

Figure 3.1 Two Measures of Similarity (Height and Sex)

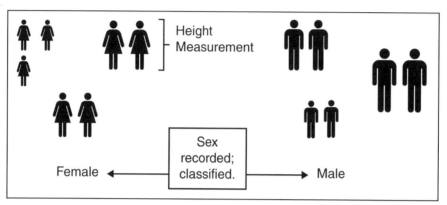

Now that both characteristics are measured or recorded, they can be classified according to a set of criteria. Here, the measure of similarity is which sex an individual is classified as, and the physical height of the individual. The concept that is taking place in this example is really a measure of *association* between the individuals on two characteristics, sex and height. Notice in Figure 3.1 that the height of the individuals forms three distinct groups: small, medium, and large heights. It is somewhat intuitive that the three groups all share something in common within each group; the members have similar heights and they are of the same sex. For practical purposes, the definitions for similarity, association, and distance are all considered synonymous. These techniques will form how we will measure the distance or association between records of customers or prospects in a database. There are a few caveats, however, that we will need to consider.

3.2 Distance Metrics As a Measure of Similarity and Association

How can we measure distance between records on a database on differing variables with different scales and have differing meanings? To demonstrate this concept a little further, consider a database of attributes as shown in Table 3.1. The fields (columns) in the database have various types (numeric, character) and scales (e.g. binary, ordinal, nominal, and interval), and they have different units.

Table 3.1 Database Field Descriptions with Differing Attributes

Field Name/Description	Measurement Type	Scale	Units
Last fiscal year revenue	Numeric	Interval	$
Filed a tax return on time (Y/N)	Character	Binary	None
Responded to direct mail (1/0)	Numeric	Binary	None
Year company was founded	Numeric	Ordinal	Years
Credit rating score (1–6)	Numeric	Ordinal	None
Industry group code	Character	Nominal	None
Distance to nearest major metro area	Numeric	Interval	Miles

If we were to measure the distance between customer records based on the variables in Table 3.1, what would be the unit of measurement when combining dollars, years, miles, industry code, and yes or no? In addition, a small change in last fiscal year revenue is not the same as a small change in miles (distance to the nearest major metro). We must translate the general concept of association into some sort of numeric measure that depicts the degree of similarity as measured by distance. The most common method, but not the only one available, is to translate all of the fields under consideration into numeric values so that the records may be treated as points in space. This is desirable because the distance between points in space can be measured from basic Euclidean geometry and a little vector algebra. It is the concept that items that are closer together distance-wise are more similar than items that are farther away from each other.

Let's take a simple example to demonstrate how distances can be measured on three rows with two fields in a database, one called AGE and one called VALUE. Table 3.2 shows the three records of AGE and VALUE. So, let's construct some distance measurements from this data set in Table 3.2. To compute distances for the AGE variable, each row must be compared with every other row along with itself. The same will be true for the VALUE variable as well. The distance measurements for AGE starting with row 1 are 8 – 8 = 0, 8 – 3 = 5, and 8 – 1 = 7. Now these are compared with the first row as the reference point. If we increment by one row and do the same, we end up with the following for the AGE variable: Row 2: 3 – 3 = 0, 3 – 1 = 2, 3 – 8 = –5 and for the last row, Row 3: 1 – 1 = 0, 1 – 3 = –2, and 1 – 8 = –7. This completes each of the distance measurements for AGE. We can do similar measurements for the VALUE variable as well, but you probably get the idea.

We can take these measurements as just described and place them in a matrix. If each value of AGE is placed in a row and the same set of values are also placed in a column, then we can construct a distance matrix for the AGE variable. Table 3.3 shows such a matrix for the AGE variable that is shown in Table 3.2.

Table 3.2 Simple Three-Row Database of Age and Value

Row #	Customer ID	Age	Value
1	Cust_A	8	$18.50
2	Cust_B	3	$3.30
3	Cust_C	1	$9.75

Table 3.3 Simple Distance Matrix for Age Variable

Row #	Cust_A	Cust_B	Cust_C
1 Cust_A	8 – 8 = 0	3 – 8 = –5	1 – 8 = –7
2 Cust_B	8 – 3 = 5	3 – 3 = 0	1 – 3 = –2
3 Cust_C	8 – 1 = 7	3 – 1 = 2	1 – 1 = 0

The matrix in Table 3.3 is symmetrical about the diagonal zeros. The upper half of the triangular portion of the matrix is identical to the lower half with the exception of a minus sign. We can compute distances of both the AGE and the VALUE fields using the SAS DISTANCE procedure. This procedure will compute Euclidian distances (as well as other types of distances too). So, let's create the data set in Table 3.2, compute the distances of AGE and both AGE and VALUE combined, see what matrix is produced, and see what these distances look like plotted in two dimensions on a graph.

Open SAS Enterprise Miner and open the project we did in Chapter 2, the Data Assay Profile project. Open the Program Editor window and place the code from the CD folder Chapter 3 called distance_example.sas. Here is the SAS code:

```
data work.age_value;
   input cust_id $ age value;
   cards;
Cust_A  8  18.50
Cust_B  3   3.30
Cust_C  1   9.75
;
run;

title 'Distances of Just the AGE Variable';
proc distance data=age_value out=age_dist method=Euclid;
   var interval(age);
id cust_id;
run;
title;
title 'Compute the Distnaces of AGE & VALUE Combined';
proc distance data=age_value out=Dist method=Euclid;
      var interval(age value);
      id cust_id;
   run;
title;
proc mds data=dist level=absolute out=mds_out;
id cust_id;
run;
%let plotitop =   gopts   = gsfmode = replace
                     device = gif
                     gaccess = gsasfile
                     hsize   = 5.63      vsize  = 3.5
                     cback   = white,
                     color   = black,
                     colors  = black,
                     options = noclip border expand, post=c:\temp\distance.gif;
%plotit(data=mds_out,datatype=mds, labelvar=cust_id, vtoh=1.75);
```

The first part of the SAS code creates a three-row data set called AGE_VALUE and puts it into the library called Work. Next, the DISTANCE procedure is run just on the AGE variable. This then creates a matrix data set called AGE_DIST in the Work library. This should be the lower half of Table 3.3. If you click the icon in the upper right corner of the SAS Enterprise Miner project window, it will open a window of SAS libraries. Select the Work library and you should now see the data set called AGE_DIST. Select the Browse option and highlight this data set. The matrix is shown in Table 3.4. The second call of the DISTANCE procedure computes the distance of both AGE and VALUE combined. Table 3.5 shows the distances of the DIST data set. If we wanted to see these distances graphically, from Table 3.5 how would we plot these values? The MDS procedure code shows the PROC MDS code and the %PLOTIT macro.

Table 3.4 Distance Matrix Computed using the SAS DISTANCE Procedure

	CUST_ID	CUST_A	CUST_B	CUST_C
1	Cust_A	0	.	.
2	Cust_B	5	0	.
3	Cust_C	7	2	0

WORK.AGE_DIST - Rows 1 - 3

Table 3.5 Distance Matrix of Both AGE and VALUE Fields

	CUST_ID	CUST_A	CUST_B	CUST_C
1	Cust_A	0	.	.
2	Cust_B	16.00125	0	.
3	Cust_C	11.20547	6.752962	0

WORK.DIST - Rows 1 - 3

PROC MDS and the %PLOTIT macro allow us to visualize these distances on a two-dimensional plane by scaling the distances of customers A through C. PROC MDS is a multi-dimensional scaling routine, which enables techniques that estimate the coordinates of a set of points that come from measuring distances between pairs of objects. The %PLOTIT macro creates a graph file in GIF format located in a folder called c:\temp\distance.GIF. This is shown in Figure 3.2. The plot shows how far each of the three customers is from each other using both AGE and VALUE together.

Now, let us consider distance and similarity computations in a little more depth.

Figure 3.2 Distance Plot of Data in Table 3.5

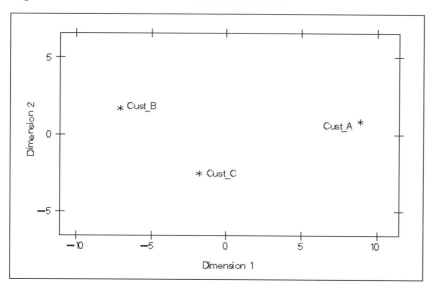

Figure 3.3 shows how distances are measured from points in a simple X-Y plane. Points X and Y are two data points. With a little help from linear algebra and geometry, we will now review some of the formulations to measure distances. The distance from point O (the origin) to point B is $|x|\cos\alpha$ and is well known from elementary geometry. This quantity is also known as the orthogonal projection of X onto Y. The points in this plane can be described as a vector from one point to another. In terms of linear algebra, the two vectors X and Y can be described as the inner product (or scalar product) and is given by Equation 3.1. The inner product of a vector with itself has a special meaning and is denoted as $X^{T}X$, and is known as the sum of squares for X. The square root of the sum of squares is the Euclidean norm or length of the vector and is written as $|X|$ or $\|X\|$.

Figure 3.3 Illustration of Distance Measurement from Inner Product

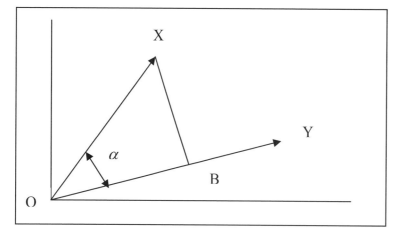

$$\langle X, Y \rangle = X^{T}Y = \sum_{i=1}^{n} x_{i}y_{i}$$

(3.1)

With this kind of notation, another way of expressing the inner product between X and Y is given by Equation 3.2.

$$X^T Y = |X||Y| \cos \alpha \tag{3.2}$$

Now if we solve Equation 3.2 for the cosine of the angles between X and Y, we get:

$$A(X,Y) = \cos \alpha = \frac{X^T Y}{|X||Y|} = \frac{\displaystyle\sum_{i=1}^{i=n} x_i y_i}{\left(\left[\displaystyle\sum_{i=1}^{i=n} x_i^2\right]\left[\displaystyle\sum_{i=1}^{i=n} y_i^2\right]\right)^{\frac{1}{2}}} \tag{3.3}$$

The cosine of the angle between X and Y is a measure of *similarity* between X and Y. So how do these formulations become important? First, remember that when there are many differing fields on your data set they must be *transformed* onto a numeric scale that has the same meaning for all the fields being considered. Second, once in the *transformed* space, the distances between each of the records can be recorded using the preceding formulas, plus a few others that are not mentioned here; see Anderberg (1973) and Duda, Hart, and Stork (2001) for other formulas. Figure 3.3 can be visualized as having the points from your database as in the *transformed* space. Then in order to understand the resulting distances, these *transformed* values can be brought back into their original dimensions and scales which will have meaning. We will study this in more detail in Chapters 5 and 6.

There have been many types of distance metrics created for special purposes. Distance metrics are especially suited for textual data, while others are designed for use with certain types of data such as binary variables or categorical variables. There are dozens if not hundreds of published techniques for measuring the similarity of data records in a table or database. The basic classes of variables or fields in your database can be nicely put into the following four groups, although others groups or classes can and do exist as well:

- categorical (also called nominal)
- ordinal or ranks
- intervals
- intervals with an origin (also called ratio)

Categorical variables give us a classification system in which to place several unordered categories to which an item belongs. We can say that a store belongs to the northern region or the western region, but we cannot say that one is greater than the other is, or judge between the stores, only that they are located in these regions. In mathematical terms, we can say that $X \neq Y$ but not whether $X \leq Y$ or $X \geq Y$. Categorical measurements, then, denote that there is a difference in type between one or more levels versus another but these measurements are not able to quantify that difference.

Ordinal or rank variables indicate to us that items have a certain specific order, but do not tell us how much difference there is between one item and another. Customers could be ranked as 1, 2, and 3, indicating that 3 is most valuable, 2 is next valuable, and 1 is least valuable, but only on relative terms. Ordinal measurements carry a lot more information than categorical measurements do. The ranking of categories should always be done subject to a particular condition. This is typically called *transitivity.* Transitivity means that if item A is ranked higher than B, and B higher than C, then A must be ranked higher than C. So, $A > B$, and $B > C$, then $A > C$.

Interval variables allow us to measure distances between two observations. If you were told that in Boston it is 48° and in southern New Hampshire it is 41°, then you would know that Boston is 7 degrees warmer than it is in New Hampshire. A special kind of interval measurement is called a *ratio* scale. *Ratios* are not dimensional in nature; that is, they don't have a set of units that goes along with the measurement value. If a ratio is based from dividing the starting speed with an ending speed, then the units of speed (e.g. miles per hour) cancel and the unit is just a ratio without dimensions.

Intervals with an origin are what some call *true measure or ratio scale.* They are considered true because they have an origin as a proper reference system. Therefore, variables like age, weight, length, volume, and the like are *true* interval measures as they have a reference point of origin that is meaningful for comparison (Berry and Linoff 1997, pp. 188–189; Pyle 1999).

Geometric distance measures are well suited for interval variables and ones with an origin. In order to use categorical variables and rankings or ordinal measures, one needs to transform them into interval variables. This can be done in a variety of ways. When we get into the algorithms of clustering (*k*-means and its variants) in Section 3.4, we will discuss how SAS performs these transformations. There is a natural loss of information as one goes from interval with an origin, to ordinals like seniority, to categories like red or blue; there is a loss of information at each stage. This should be remembered when converting variables such as age into ranks.

What we have seen so far is that vectors can represent points in a geometric plane, and the distances and the relative association between the vectors can be represented mathematically. As said earlier, there are many published techniques for measuring the similarity of records in a database. For our purposes, the distance between two points is used as the measure of association. In this scenario, each field in a row or record becomes one element in a vector that describes a point in space. If the points are close to each other distance wise, the respective records in the database are also considered similar for that feature. There are also many metrics that can be used to measure the distance between two points, but the most common one is the Euclidian distance.

In terms of mathematics, Table 3.6 gives some formal definitions of distance. Any function that takes two or more points and produces a single number describing a relationship between the points is a potential candidate for measure of association; however, a true distance metric must follow the rules in Table 3.6 (Berry and Linoff 1997, pp. 188–189).

Table 3.6 Distance Metrics Defined

$D(X,Y) = 0$ if and only if X=Y ❶

$D(X,Y) \geq 0$ for all X and all Y ❷

$D(X,Y) = D(Y,X)$ ❸

$D(X,Y) \leq D(X,Z) + D(Z,Y)$ ❹

❶ This property implies that if the distance is zero, then both points must be identical.

❷ This property states that all distances must be positive. Vectors have both magnitude and direction; however, a distance between the vectors or points must be a positive number.

❸ This property ensures symmetry by requiring the distance from X to Y to be the same as the distance from Y to X. In non-Euclidean geometry, this property does not necessarily hold true.

❹ This property is known as the triangle inequality and it requires that the length of one side of a triangle be no longer than the sum of the lengths of the other two sides (Anderberg 1973).

An example might be helpful here. Table 3.7 shows seven records in a sales customer database. The fields are age of the person, revenue of items purchased, and state where the individual resides.

Table 3.7 Seven Customer Records in a Customer Database

Row Number	Customer ID	Age (years)	Revenue ($)	State
1	372185321	28	$155	CA
2	075189457	55	$68	WA
3	538590043	32	$164	OH
4	112785896	40	$596	PA
5	678408574	26	$48	ME
6	009873687	45	$320	KS
7	138569322	37	$190	FL

To compute the distance between row 1 and 2 for age is straightforward. Distance(age)[1,2] = abs(55 − 28) = 27. However, if we want to compare this distance to rows 1 and 2 for revenue, Distance(revenue)[1,2] = abs(68 − 155) = 87 is not the same set of units. We need to transform these so that they are on the same relative scale; a scale between 0 and 1 would be one possible choice. We can do this by taking the absolute value of the difference and then dividing by the maximum difference. The maximum difference in age is the maximum − minimum; in age the max(age) is 55 and the min(age) is 26. Then, the normalized absolute value difference in age for rows 1 and 2 now is: Distance(age)[1,2] = abs(28 − 55) / (55 − 26) = 27 / 29 = 0.93103. The same kind of computations can be done for revenue as well. Distance(revenue)[1,2] = abs(155 − 68) / (596 − 48) = 87 / 548 = 0.1587. State is a categorical variable and one method of transforming this is to transpose the state so that each unique level of state is a separate dummy variable for each level of state. This is shown in Table 3.8.

Table 3.8 State Variable Dummy Transformations

State	State - CA	State - WA	State - OH	State - PA	State - ME	State - KS	State - FL
CA	1	0	0	0	0	0	0
WA	0	1	0	0	0	0	0
OH	0	0	1	0	0	0	0
PA	0	0	0	1	0	0	0
ME	0	0	0	0	1	0	0
KS	0	0	0	0	0	1	0
FL	0	0	0	0	0	0	1

This set of dummy variables allows the distances of a categorical variable like State and transforms it so that distances are measured on a scale between 0 and 1. These are not distances in miles between states, but likeness of records in the database to have a similar state name. So, if we now compute all the Euclidean distances and normalize them as shown earlier, Table 3.9 shows the matrix of normalized distances for the variable Age on the seven database records in Table 3.7.

Table 3.9 Normalized Distance Metrics of Age from Table 3.7

Customer ID	37218532	07518945	53859004	11278589	67840857	00987368	13856932
37218532	0	.	.	.	.	.	.
07518945	4.60969	0	.	.	.	.	.
53859004	3.76254	4.40064	0	.	.	.	.
11278589	4.57006	4.90402	4.45998	0	.	.	.
67840857	3.78974	4.704	3.83767	4.93738	0	.	.
00987368	4.19027	4.09385	4.03964	4.04894	4.42432	0	.
13856932	3.8492	4.18897	3.77629	4.32954	3.96706	3.88526	0

The same sort of normalized distance metrics can also be applied to the revenue field and the state field when the states are set up with dummy variables as in Table 3.2.

For interval data, a general class of distance metrics for n-dimensional patterns is called the Minkowski metric and is expressed in the form of Equation 3.4.

$$D_p(X_j, X_k) = \left(\sum_{i=1}^{i=n} \left| X_{ij} - X_{ik} \right|^p \right)^{1/p}$$

(3.4)

This metric is also known as the L_p norm. When p is 1, the metric is called the *city-block* or Manhattan distance, when p is 2, the Euclidean distance is obtained, and when p is 3, the Chebychev metric is derived (Anderberg 1973; Duda, Hart, and Stork 2001). So, the distances in Table 3.9 were of the form when $p = 1$, and the values were normalized by dividing by the maximum less the minimum value. Many other derivations can be obtained in this fashion depending on the overall objective. If you are getting the feeling that distance metrics are rather compute intensive, you're right. They are. When these computations are done on many records and many fields they are very compute intensive.

3.3 What Is Clustering? The *k*-Means Algorithm and Variations

If you refer back to the Hertzsprung-Russell diagram (Figure 1.4) in Chapter 1, "Introduction," the luminosity and temperature plot of stars produces natural clusters of various stages in the life-cycle of a star. We've discussed how one can measure the degree of similarity by measuring distances; then items that are closer in distance are more like each other than items that are farther apart. The Hertzsprung-Russell diagram is a simple example that has a meaningful geometric representation that can be visualized in two dimensions. What happens when we have many fields in a database and thus many dimensions of distances to cluster? As the number of dimensions increases, the ability to visualize clusters and use our intuition about the distances quickly becomes a daunting task that is often not feasible. This is where an algorithm is needed. The first to coin the term *k-means* was J. B. MacQueen, (1967, pp. 281–297) who used this term to denote the process of assigning each data element to that cluster (of k clusters) with the nearest centroid (mean). The key part of the algorithm is that the cluster centroid is computed on the basis of the cluster's current membership rather than its membership at the end of the last cycle of computations as other methods have done (Anderberg 1973). A *cluster* is nothing more than a group of database records that have something measurable in common; however, the basic structure of the groups is not known or defined. When a reference to a clustering algorithm is given, the reference is usually meaning an algorithm that is *undirected* as pointed out in Section 3.1.

Currently, the *k*-means method of cluster detection is one of the most widely used in practice. It also has quite a few variations. The *k*-means method was the main one that sparked the primary use in SAS/STAT, the FASTCLUS procedure. The selection of the number of clusters, k, has often been glossed over because the loop in the algorithm that selects a different k is really the analyst and not the computer program. What is typically done is after one selects a value of k, the resulting clusters are evaluated, then trying again with a different value of k. After each iteration, the strength of the resulting clusters can be evaluated by comparing the average distances between records in a cluster with the average distance between clusters, and there other methods as well as discussed later in this section. However, this kind of iteration could be performed by the program, but an even more important issue arises in the cluster evaluation and that is the overall usefulness of the resulting clusters. Even well separated and clearly defined clusters, if not useful to the analyst or the desired application, have very little purpose in business or industry (Berry and Linoff 1997, 99. 188–189). The *k*-means algorithm is simple enough to specify (Duda, Hart, and Stork 2001):

Algorithm 1 (k-means clustering)

begin: initialize n, k and $u_1, u_2, u_3, \ldots u_k$

classify n samples according to the nearest μ_i

recompute μ_i

until no change in μ_i

return the values of $u_1, u_2, u_3, \ldots u_k$

end:

where μ_i is the mean, n is the number of samples,

and k is the number of clusters.

Typically, the *real* value of k is sometimes denoted as *c*, and then the estimated value of *c* is denoted as k. By *real* I mean the actual number of true clusters (if they exist) in the data set at hand, and *c* would be the estimate of k.

In geometry, all dimensions are equally important. As said earlier, what if certain fields in our database are measured in differing units like that indicated in Table 3.7? These units must all be converted to the same *scale*. In Table 3.7, we cannot use one set of units, say dollars for revenue, and try to convert age to dollars. The solution then is to map all the variables to a common *range* (like 0 to 1 or –1 to 1 or 0 to 100, etc.). That way, at least the ratios of change of one variable are comparable to the change in another variable. I refer this remapping to a common range as *scaling* (Berry and Linoff 1997, pp. 188–189). The following list shows several methods for scaling variables to bring them into comparable ranges:

- Divide each variable by the mean (e.g. each entry in a field is divided by the mean of the entire field).

- Subtract the mean value from each field and then divide by the standard deviation. In statistical terms, this is called a *z* score.

- Divide each field by the range (difference between the highest and lowest value) after subtracting the lowest value.

- Create a normal scale by the following equation:

$$V_{norm} = \frac{V_i - \min(V_1...V_n)}{\max(V_1...V_n) - \min(V_1...V_n)}$$

Scaling takes care of the problem wherein changes in one variable appear more significant than changes in another because of the rate at which the units they are measured in get incremented. Referring to Table 3.7, what if revenue is more important to us than age and we want to take that into consideration in the clustering algorithm? This kind of issue calls for a modification of weights so that variables that are more important carry a larger weight than variables that are less important. When you change a weighting scheme to your algorithm, you also add an additional criterion for each iteration in the cluster computations; you will want to evaluate the effects of various weighting strategies to see if the weights have produced a desired result. We will discuss more on scaling of variables in Chapter 6, "Clustering of Many Attributes."

3.3.1 Variations of the *k*-Means Algorithm

The general form of the *k*-means clustering has a lot of variations. There are methods of selecting the initial seeds of the clusters, or methods of computing the next centroid, or using probability density rather than distance to associate records with the clusters. The *k*-means clustering has the following drawbacks:

- It does not behave well when there are overlapping clusters.

- The cluster centers can be shifted due to outliers. (We will discuss how to deal with some of these later.)

- Each record is either in or not in a cluster, although un-clustered outliers can be reviewed later. There is no notion of probability of cluster membership; e.g., this record has an 80% likelihood of being in cluster 1.

Fuzzy *k*-means clustering has been developed to simulate the third item in the preceding list of issues from classical *k*-means clustering. The *fuzzy* cluster membership is a probability that a database record belongs to a particular cluster or even to several clusters. The probability distribution often used is a Gaussian distribution (e.g., a normal bell-shaped curve distribution). These variants of *k*-means are called Gaussian mixture models. Their name comes from a probability distribution assumed for highly dimensional type of problems. The seeds are now the mean of a Gaussian distribution. During the estimation portion, this type of fuzzy membership is depicted in Table 3.6. The darker cluster members have a probability of membership greater than 80%, the lighter cluster members have a probability of membership less than 80% but greater than 60%, and the lightest group of cluster members have less than 60% but greater than 40% probability of cluster membership. Although the lighter elements have a lower probability of being a member of the darker elements, they could have a high probability of being a member of another cluster. During the maximization step of the fuzzy algorithm, the association or responsibility that each Gaussian has for each data point will be used as weights immediately following the maximization step. Each Gaussian is shown as a G_1, G_2, G_n in Figure 3.4. Fuzzy clustering is sometimes referred to as soft clustering.

Figure 3.4 Illustration of Fuzzy Cluster Membership

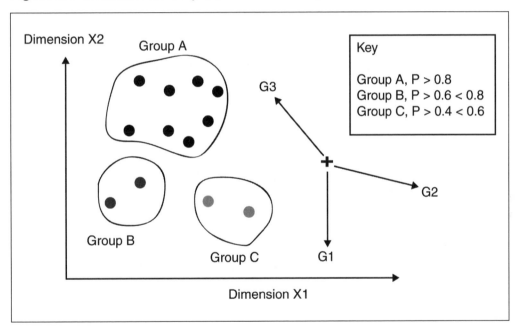

3.3.2 The Agglomerative Algorithm

There are different types of clustering and many algorithms to choose from a rather long list. This section is intended to give you a flavor for the types of clustering methods. Most clustering techniques can be placed into two types, *disjoint or hierarchical.* Disjoint does not refer to bones that are out of socket; it refers to clusters that don't overlap, and each record in the database belongs to only one cluster (or perhaps an outlier that does not belong to any particular cluster). Hierarchical clusters are ones in which a data record could belong to more than one cluster and a hierarchical tree can be formulated that describes the clusters. This is particularly useful when building a taxonomy or trying to understand the possible structure in the data that may otherwise be unknown. Such a tree that shows this hierarchy is typically called a dendrogram, and an example of one is shown in Figure 3.5. The points X1–X8 are individual records, and the horizontal axis represents the generalized measure of similarity among the clusters. Points X2, X3, X6, and X7 happen to be very similar and are merged at the 2nd level. X4 and X5 are similar and are merged at the 3rd level, etc. (Duda, Hart, and Stork 2001.)

In order for the agglomerative technique to work, it must make a *similarity matrix*. This is similar to the one shown in Table 3.7. If you will notice in Table 3.7, only half of the information is really needed as the half above or below the zero diagonals are the exact same distance measurements. This is true because of the third property in Figure 3.4 that states that the distance from points A to B is identical to points B to A. The general agglomerative algorithm looks like the following:

1. Start with *n* clusters each consisting of exactly one entry or record. Label the clusters 1 through *n*.
2. Look through the similarity matrix for the most similar pair of clusters. Label the chosen similar clusters *p* and *q*.
3. Merge clusters *p* and *q*, reduce the total number of clusters by *n-1*, and update the similarity matrix.
4. Perform steps 2–3 for a total of *n–1* times after which all the records belong to the same large cluster. At each iteration, record which clusters were merged and how far apart they were.

Figure 3.5 Hypothetical Example of Dendogram Forming Hierarchical Clusters

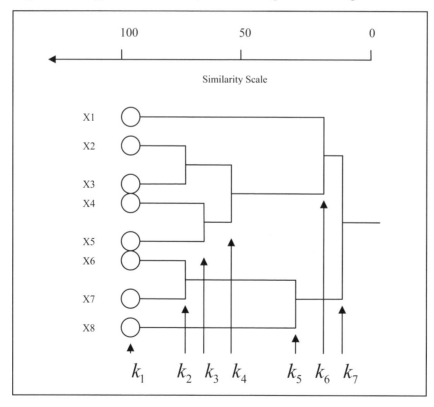

From the simple customer table in Table 3.7 and the coding of states as in Table 3.8, a simple clustering produces a simple dendogram of hierarchical clusters as in Figure 3.6.

Figure 3.6 Hierarchical Clusters in Simple Customer IDs in Table 3.7

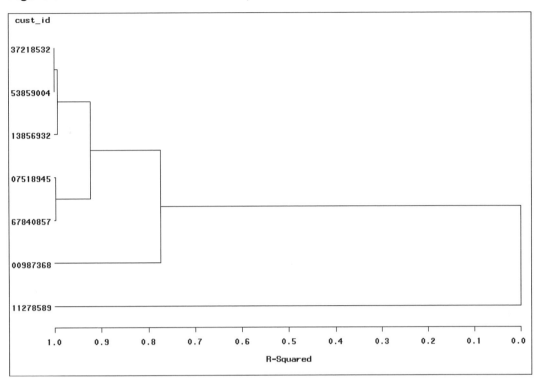

The information obtained from recording how far apart the clusters are will be useful, as we now need to make a determination of how to measure the distance between the clusters. In the first iteration through the cluster merge step, the clusters to be merged each contain only one entry or record so the distance between clusters is the same as the distance between the records. However, on the second pass through and in subsequent passes, we need to update the similarity matrix with the distances from the multi-record cluster to all of the other clusters. Again, there are choices to make on how we can measure the distances between the clusters. Here are the three most common approaches:

- single linkage
- complete linkage
- difference or comparison of centroids

In the single linkage method, the distance between any of the clusters is determined by the distance between the closest members. This method produces clusters so that each member of a cluster is more closely related to each other than any other point outside that cluster; i.e., it will tend to find clusters that are more dense and closer to each other than in the other methods. In the complete linkage method, the distance between any of the clusters is given by the distance between their most distant members. This produces clusters with the property that all members lie within some known maximum distance of one another. Moreover, in the third method of centroids, the distance between the clusters is measured between the centroids of each (the centroids is the average or mean of the elements) (Berry and Linoff 1997, pp. 188–189; Anderberg 1973). Figure 3.7 shows a representation of the three methods (Berry and Linoff 1997, pp. 188–189).

Figure 3.7 Three Common Methods for Measuring Cluster Distances

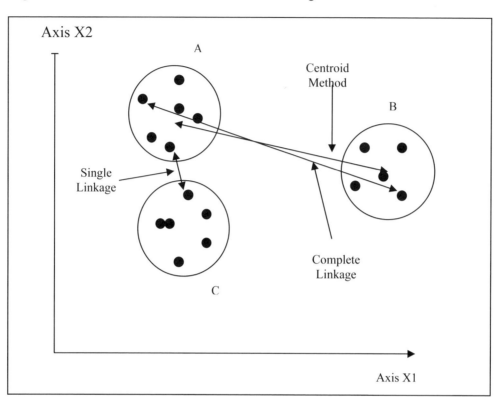

In the *k*-means approach to cluster analysis, we need a way to figure out what value of k determines the best clusters. In a similar fashion, when performing hierarchical clustering, we need a method to test which level in the hierarchy contains the best clusters. However, what criteria do we use to determine *good* clusters? When is a cluster good or rather good enough? For most CRM applications, good typically refers to the customer records in each cluster that are similar to each other by the criteria we selected for the clustering, but in general terms we want clusters whose members are very *similar* to each other while at the same time the clusters themselves are well separated. The farther the cluster separation, the greater the differences in customer attributes that each cluster represents. Referring to Figure 3.7, this means that customers in cluster A are all like each other in some fashion (e.g., they mostly purchase through an indirect purchase channel), in cluster B they purchase only by a direct channels, and perhaps in cluster C those customers purchase in both direct and indirect channels. A typical measure of the within-cluster similarity is the variance (the sum of the squared differences of each element from the mean). A general rule-of-thumb that works for both disjoint and hierarchical clustering is to use whatever similarity measure or distance was used to form the clusters and also use this to compare the average distance between the clusters (Berry and Linoff 1997, pp. 188–189). When a set of scoring code is used to project a clustering algorithm on a much larger data set, one way to evaluate if a re-clustering is needed is to look at the member distances from their mean in each cluster, and if they have moved substantially from the previous clustering, then this may be a good indication that you need to re-cluster the original data set. We will use this in more detail in Chapter 7, "When and How to Update Cluster Segments," when we review the shelf life of a clustering model.

3.4 References

Anderberg, Michael R. 1973. *Cluster Analysis for Applications*. New York and London: Academic Press.

Berry, Michael J. A., and Gordon S. Linoff. 1997. *Data Mining Techniques: for Marketing, Sales, and Customer Support*. New York: John Wiley & Sons.

Duda, Richard O., Peter E. Hart, and David G. Stork. 2001. *Pattern Classification*. 2d ed. New York: John Wiley & Sons, Inc.

MacQueen, J. B. 1967. "Some Methods for Classification and Analysis of Multivariate Observations." *Proceedings of the Berkley Symposium on Mathematical Statistics and Probability*. Berkley: University of California Press. 1:281–297.

Pyle, Dorian. 1999. *Data Preparation for Data Mining*. San Francisco: Morgan Kaufmann Publishers, Inc.

SAS Enterprise Miner, Release 5.2. Cary, NC: SAS Institute Inc.

Part 2

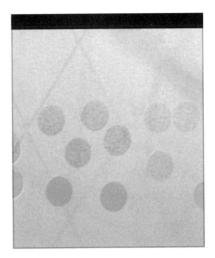

Segmentation Galore

Segmentation Using a Cell-based Approach

4.1 Introduction to Cell-based Segmentation

As indicated in Chapter 1, "Introduction," segmentation is the process of dividing up records in your database, be it customers or prospects, and somehow classifying them into specific categories, groups, or segments. The process of doing this and subsequent profiling allows a better understanding of those customers or prospects and so segmentation is often used for just that, getting to know and understand your customers or prospects. Just as we did some profiling in Chapter 2, "Why Segment? The Motivation for Segment-based Descriptive Models," segmentation can also be thought of as a type of profiling, creating segments or groups that have some like, or similar, characteristics. The eventual plans for using the segments will determine the best method or approach at creating them, so when performing segmentation, one needs to have a business plan or problem to solve prior to the segmentation analysis (Rud 2000). A business problem may come in a variety of forms. It might be a question arising from a field sales representative or a sales representative in a call or contact center. It could be a directive from upper management in the form of a goal or objective to meet in the next fiscal quarter, half, or year. But the first thing that one should do in a business or industrial setting is to define the objectives and goals for the business problem at hand.

One way to classify customers is by devising a segmentation method that combines attributes that are desirable and then performing this method on the entire database. This technique is often referred to as *scoring*. Scoring customers according to one or a few of several attributes typically constitutes a *cell* much like a cross section in a matrix. The intersection of a row and column in a matrix represents the *i*th attribute in a row and the *j*th attribute of a column as illustrated in Table 4.1.

Table 4.1 Matrix Representation of Patient Attributes

In Health Care Data
Columns Representing Age – *j*

	HR	10 – 20 yrs	20 – 30 yrs	30 – 40 yrs	40 – 50 yrs
Rows Representing Heart Rate - *i*	**60 bpm**	A	B	C	D
	65 bpm	E	F	G	H
	70 bpm	I	J	K	L
	75 bpm	M	N	O	P

The columns in Table 4.1 denoted as *j* are deciles of age in years, and the rows denoted as *i* are 5 beats per minute (bpm) increments each. Therefore, cell G is a cross section that represents all patients that are between the ages of 30–40 and have 65 bpm of heart rate. Table 4.1 is a segmentation of two physical health care attributes of patients in a patient database, for example. Knowing the counts and percentages of each cell label in Table 4.1 allows a brief profile between both of the physical attributes of patients using a simple one-way frequency distribution table. If there is a normal distribution of heart rates from 60–75 bpm and if there is a normal distribution of ages from 10–50 years, then one would expect that the bulk of patients would fall in cells F, G, J, and K. Cells A, D, M, and P would be *outlier cells*, meaning they would represent the more extreme tails of both heart rate and age. Now the accuracy of these cells is dependent on the measurements of heart rate and the classification system. For example, the difference between patients classified into cells A and E could be accounted for +/– 5 years of age and +/– 2½ beats per minute in each cell. If your apparatus for measuring heart rate is within 0.5 beats per minute, then the classification by increments of 5 beats per minutes could be off by about 0.5. The main point of this is you now have a simple set of classes that represent two physical attributes on your database with which to score, analyze, segment, and manage.

4.2 Segmentation Using Cell Groups—RFM

One of the most common types of segment profiles used in direct marketing originated in the catalog industry is called recency, frequency, and monetary value (RFM). Just like the classes used in Table 4.1, imagine a three-dimensional table that represents your customer with these three attributes.

Recency

This attribute is how recent your customers purchased from you. It has long been known that customers who purchased recently are more likely to purchase again, compared to a customer who has not purchased in a long time. Time in this case can mean anything from days, months, quarters, years, or whatever is useful in your particular line of business or industry. Because the kind of recency greatly depends on the type of items being purchased and the line of business you're in, the level of recency segments will need to be scrutinized carefully by business managers, consultants, and the like who know from experience how often is often enough, etc. For example, in the computer and technology industry, purchasing a handheld PDA device or software would be purchased at very different intervals from a high-end UNIX-based server. In the automobile industry, a person on average might purchase a new or used car every four to six years. Whatever the typical buying cycle of the product or service portfolio you are offering in your business, a recency computation will be valuable to segment your customers and test the idea that more recent customers are more likely to purchase than not-so-recent customers. Pay careful attention to the fact that these recency cells are not necessarily predictive, but they should be tested to see if they are predictive. Intuition and experience tells us that a more recent customer is more likely to purchase than an old customer who hasn't purchased in recent times, but this should be tested periodically to see if this holds true in your business and within the database of your customers.

Frequency

Frequency is how much a customer purchases from your business and the number of purchases. Combining frequency with RFM allows for differences in volume purchases levels obtained from customer transactions. If a customer purchases once in 12 months and another customer purchases eight times in the same time period, then one can say that even if the amount of revenue or profit is identical, the customer who purchases more often is more likely to continue purchasing and is more likely to be a loyal customer than the customer who purchases only once for the same or even higher value. As a general rule, higher customer loyalty and value is usually obtained from customers who purchase more frequently than who purchase less frequently in similar time frames.

The attribute of frequency is an indicator of your customers' demand level. One way to think of this is to imagine you are a store owner and you observe the number of times a particular customer walks into your store to purchase something. The more times the customer engages with you, the more opportunities you have to perhaps cross sell items that the customer may not have, etc. Again, more opportunity to engage means more potential, but the opportunity demands action; otherwise, the frequency of times could drop less.

Monetary Value

This means just that, value in either revenue, profit, or some derived computation thereof. Similar to frequency, it can be during a specific time frame or can include all purchases made by a specific customer. The attribute of dollar amount is typically less predictive of the two in an RFM model. Yes, that's right, model; RFM is a model that incorporates three customer attributes of their purchases into one unit or cell. Monetary value can be a computed or derived value such as lifetime value (LV) or more specifically for each customer, customer lifetime value (CLTV). I will discuss estimating lifetime value in Section 8.4.

This next example of monetary value may seem very basic; however, many marketing and sales professionals do not typically compare their customers using these techniques as they should. Customers are different and if one is going to have a customer-based marketing and sales effort, knowledge about the customer's behavior is paramount. Let us examine the customer attributes from a database that has two customers in our study, Customers A and B. Assume for the moment that the inflation rate has been at 3% and, for all practical purposes, we will assume it to be constant for this example. Customer A has purchased four items of one quantity each in the last year from your business, and Customer B has purchased three items also of one quantity each in the same time frame. These two customers and their purchases are shown in Table 4.2. The total net revenue for all items is shown in the sixth column, and a computation of the net present value with a rate of 3% for one year is computed in the last column. Customer A has a slightly higher net present value (NPV) because of item 3 purchased.

Table 4.2 Customers A and B Purchases in One Year

Customer	Item 1	Item 2	Item 3	Item 4	Net Total Revenue	NPV rate of 3%
A	$50.75	$500.35	$75.00	$1200.00	$1826.10	**$1,772.91**
B	$50.75	$500.35	$0.00	$1200.00	$1751.10	**$1,700.10**

The difference between the NPV values of Customer A and Customer B is slightly less than that of the actual net revenues: net revenues of Customer A less Customer B is $75.00, whereas NPV of Customer A versus B is $72.81. This is because the time value of money has been taken into account. The monetary component of RFM can be the total net revenue or the net present value of revenue or perhaps a future value of revenue. Computing NPV will be reviewed in further detail in Chapter 8, "Using Segments in Predictive Models." These computations given are rather simple and one would normally compute the net profit by taking into account the pricing and cost of items 1 through 4. If you again performed these computations in one more year and the items purchased did not increase in the NPV, and the inflation remained the same, then the monetary value of the customers in year two has diminished and is a sign of a critical business condition. If in year two the NPV of both customers diminished in value due to inflation and in their purchases, this is a sign of an ailing business and can be terminal for your business if not attended to. This kind of nugget of information obtained from the data of your customer database is vital to the understanding of your business. Knowledge of your customer's purchasing condition is vital; however, if you don't do anything with that knowledge in order to correct the situation, then you have actually cost the corporation. The effort and time to compute, store, validate, data cleanse, etc. does cost your company in resources. Therefore, the key to the business improvement is in the fixing of the situation once it has been assessed. We will come back to this thought in Sections 4.5 and 4.6.

Back in Chapter 1, "Introduction," Figure 1.2 depicted an RFM cell structure in the form of a cube. That graphic is replicated here in Figure 4.1 to show that customers classified in various group cells can be organized for the purpose of strategizing sales or direct marketing communications. All customers in group 3C1, for example, could be good candidates to increase revenue as they are very recent and they have purchased often. The company has increased their revenue flow by perhaps cross-selling or up-selling to those customers. A set of customers in 3C3 is the highest in monetary value and loyalty in this illustration and as such, a program should be created for them as well. These RFM cells are methods for classification schemes and are not forecasts of future behavior. This means that the RFM cells are not necessarily predictive, but they do a wonderful job of classifying three attributes of customer behavior into a single class that is straightforward to use.

Figure 4.1 RFM Cell Pictorial Description

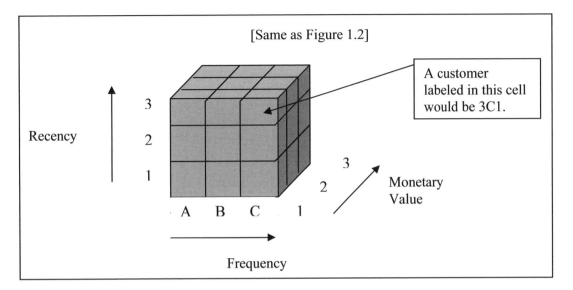

The reason or motivation for performing customer segmentation is typically based on a need to improve business performance or obtain some business objectives. This involves understanding *why* segmentation is being performed. Having knowledge of the main goal or key business objective is paramount to determining a strategy to improve performance. What is needed to perform this is a process for segmenting our customer or prospect data in order to focus our efforts. Framing the business problem is an extremely useful task in order to get the big picture of the business issue, problem, or goal that is involved. To demonstrate this, consider the following situation. Two nature enthusiasts were walking in a wetland area in southern Florida when suddenly they were faced with a couple of large crocodiles. Frozen in their tracks, one of the enthusiasts started frantically removing his backpack and equipment and the other person said "What in blazes are you doing? You can't outrun those crocks." The other person said "I don't have to outrun those crocks, I only have to outrun you." So, it's not a very funny joke; however, it does bring out perhaps a smile or two. Why? Well, at first the problem seems to be two people against hungry crocodiles and how to escape. When one of the enthusiasts re-frames the problem in a different way, this brings out somewhat of a surprise and the situation now has a whole different meaning. It is this re-framing of the problem that is many times key to segmentation for the purpose of solving a business problem or issue. So, the first key in good customer segmentation is to understand the business problem at hand, and the second is to understand how to frame or perhaps re-frame the problem (Pyle 2003).

Segments of your customers, in order to be effective, must be relevant to the business issue or problem. Customer segments may be wonderfully grouped with fairly equal sizes, statistical significance from each other, good separation, and definitive profiles, etc. However, if they are not relevant, they are of little use for solving a business issue or problem.

4.2.1 Other Cell Types for Segmentation

RFM is not the only type of segmentation cell methodology. Others can be designed to reflect some sort of business issue. If demographics are an important feature for a specific program, such as industry or company employee size, then specific demographic segments can be made to fit a particular business rule. Table 1.1 in Chapter 1 demonstrated industry segmentation using SIC codes to classify customers or prospects. Just as RFM is a cell code that combines three customer attributes, a cell code can also combine demographic attributes. Table 4.1 demonstrates two attributes of patients, age and heart rate into a single cell code. There is no end to the possibilities of creating cell code segments on your customer or prospect databases; they should be made to reflect a business rule or process that intends to use them for classification.

Life stages of a customer are also another way to group and classify these customer attributes. Whether we like it or not, each of us ages and with our aging we go through certain life stages and patterns that change over time to meet our specific needs. Customers have life stages as well. In a consumer business, life stages are often grouped into teens, singles or young couples, middle-aged families, or seniors, etc. Additional enhancements can be gained by overlaying other behavioral, financial, and psychographic data to create well-defined business segments that can be used for marketing strategies (Rud 2000).

I've spent a fair amount of time discussing the benefits and types of cell type segments so let's get into actually building RFM cells and see how they can be used in a the context of CRM.

4.3 Example Development of RFM Cells

To begin our example, start SAS Enterprise Miner and open a new project called RFM Cell Development. From the data CD, copy the BUYTEST data set to the SAMPSIO library location. The process flow table follows.

Process Flow Table: RFM Cell Development

Step	Process Step Description	Brief Rationale
1	Start SAS Enterprise Miner and create a new project.	
2	Add a data source to the data mining flow—BUYTEST data set. Use the Data Source wizard in SAS Enterprise Miner.	BUYTEST data set is explained in Appendix 1.
3	Create a new diagram in SAS Enterprise Miner called RFM Cells.	
4	Explore the VALUE24 variable in the input data source node.	Performs simple Data Assay on VALUE24.
5	Create a Transform Variables node to subdivide VALUE24 into intervals.	Breaks up VALUE24 into quartile groups.
6	Create a single variable that contains attributes of RFM using quantiles. Use a SAS Code node to write custom statements for the RFM variable.	Makes a single variable that has categorical levels that correspond to RFM cells.
7	Add a FREQUENCY procedure and format to the RFM code in the SAS Code node.	Allows checking the RFM cells just created.
8	Add a Metadata node to the diagram to change roles of variables.	Revises variable roles to create a predictive model.
9	Add a Data Partition node for splitting the data set into training, validation, and test sets for the model development.	The test set is not used in the model development but is used for scoring to see how well the model generalizes the predictions.

(continued)

Process Flow Table (*continued*)

Step	Process Step Description	Brief Rationale
10	Place a Decision Tree on the flow diagram, which will model the RESPOND variable.	Predicts the RESPOND variable using RFM cells and other variables.
11	Add a Model Comparison node for assessing lift charts of RESPOND.	Assesses the model predictions on each of the partition data sets: training, validation, and test.
12	Review the actual tree in the Decision Tree results window.	Observes the rules of the tree and other model characteristics.
13	Add a Score node to score the test data set from the partition.	Scores new data that was not used in the model development.
14	Observe the actual vs. predicted classification in Decision Tree results.	Assesses the accuracy of the predicted classifications.
15	View the results in the Model Comparison node.	Lifts chart assessment of the model.

Step 1: Click the Data Sources folder and add a new data source. Add this data source using either the basic or the advanced option in the Metadata Advisor window. **Step 2:** This data set is identical to the one we used in Chapter 2. It contains about 10,000 customer records in which we will create an RFM score. Some elements are already computed for you such as the number of purchases in six-month intervals, for example. The information we need to develop an RFM cell score is contained in four variables: BUY6, BUY12, BUY18, and VALUE24. The BUY6 through BUY18 fields describe the count of items purchase in those six-month intervals, thus giving us both recency and frequency. The monetary value is in the field VALUE24, which is the total dollar amount purchased in the last 24 months. In order to best determine how to break out the monetary value, let's see the distribution of VALUE24 and look for some clues for natural break points. To do this, run PROC UNIVARIATE on the BUYTEST data set using the following statements.

Step 3: Create a new diagram and call it RFM Cells. Drag the BUYTEST data source to the flow diagram and when you highlight the Data node, you should be able to click the **Variables** property sheet icon in the Properties window. **Step 4:** Highlight the VALUE24 variable and click the **Explore** button. This should open a window with a histogram of the distribution of values. Figure 4.2 shows what that distribution should look like. What we would like to do is to break up this distribution into quartiles for further processing.

Figure 4.2 Distribution Output of VALUE24 Field on BUYTEST Data Set

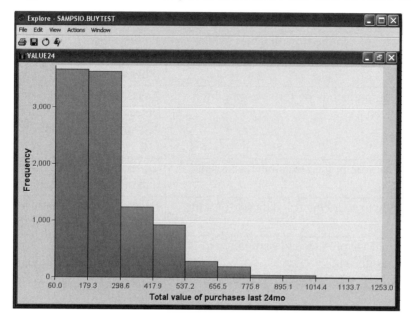

Although there are a number of ways in which we can perform these computations, we will use a combination of the Transform Variables node and the SAS Code node. **Step 5:** Drag a Transform Variables node and connect the BUYTEST data node to it. Click the property sheet labeled **Variables**, which opens a window allowing you to add standard formulas used to transform variables. Highlight the variable VALUE24 and in the **Method** selection, select **Quantile**. A quantile category is when you take the numeric distribution and break it into four evenly sized segments.

Now, drag a SAS Code node and connect the Transform Variables node to it. If you now click **Variables** in the SAS Code node property sheet, you should be able explore the new distribution in quantile categories as in Figure 4.3. Use the **Explore** button on PCTL_VALUE24 and see the newly created quantiles as in Figure 4.4.

Figure 4.3 SAS Code Node Variables Listing

ID	Yes	No	ID	Nominal	C		Customer ID
INCOME	Yes	No	Input	Interval	N		Yr Income in tl
LOC	Yes	No	Input	Nominal	C		Location of res
MARRIED	Yes	No	Input	Binary	N		1 if Married, 0
ORGSRC	Yes	No	Input	Nominal	C		Original custo
OWNHOME	Yes	No	Input	Binary	N		1 if own home
PCTL_VALUE24	Yes	No	Input	Nominal	C		Transformed:
PURCHTOT	Yes	No	Input	Interval	N		Test mailing p
RESPOND	Yes	No	Input	Binary	N		1 if responded
RETURN24	Yes	No	Input	Binary	N		1 if product ret
SEX	Yes	No	Input	Binary	C		F or M

Explore...   OK   Cancel

Figure 4.4 SAS Code Explore Variable PCTL_VALUE24

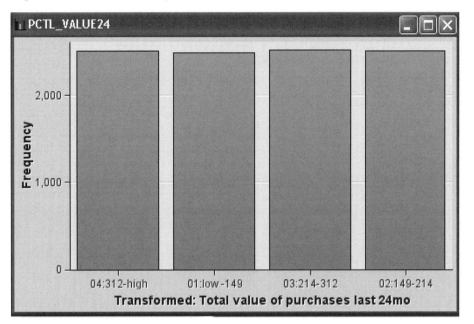

You should now have four equally sized quantiles located at the 25th, 50th, and 75th percentiles of the variable VALUE24. This could easily have been done with SAS DATA step and procedure code; however, this was intended to show how to perform these easily within SAS Enterprise Miner, which saves time and thus provides a lot of value for analyst productivity. Other types of transforms are available and even custom ones; we'll look at those in more detail when we need to transform numeric variables in Chapters 5 and 6.

Step 6: We could stop at this point with your RFM contained in a categorized form of value (the newly created quantiles variable) and some combination of the variables BUY6 through BUY18. However, let's create a single variable that contains the components of RFM and thus have a single RFM segmentation field. We'll see why this can be an important distinction later on. To do this, open the SAS Code node property sheet **SAS Code** in the Code section, open the SAS Code node, use the **Edit** menu, and select the **SAS Editor**. Now right-click in the Code section, select the **File** menu, and select the SAS code called RFM.SAS located in the Chapter 4 folder of your data CD. Your output should now look like Figure 4.5. Now, just below this DATA step code, place the following SAS FORMAT and SAS FREQUENCY procedure statements as given in Figure 4.6. The SAS FREQUENCY procedure performs many more things than simple distribution reports; however, distribution or crosstabulation reports typically need one or two lines of SAS statements.

Figure 4.5 RFM Cell Develop Project—SAS Code Node Entry

```
M SAS Code

Macro Variable          Current Value          General
General
EM_USERID               Author                 System macro variables

Macro Variables    Macros

/* RFM Cell code development - Chapt. 4 */
data &EM_EXPORT_TRAIN;
length rfm $1;
  set &EM_IMPORT_DATA;
    if (PCTL_VALUE24)='01:low -149' then do;
         if buy18=0 and buy12=0 and buy6=0 then RFM='A';
         if buy18 ge 1 or buy12 ge 1 or buy6 ge 1 then RFM='B';
           if buy6=1 and buy12=1 and buy18=1 then RFM='C';
    end;
      if (PCTL_VALUE24)='02:149-214' then do;
         if buy18=0 and buy12=0 and buy6=0 then RFM='D';
         if buy18 ge 1 or buy12 ge 1 or buy6 ge 1 then RFM='E';
           if buy6=1 and buy12=1 and buy18=1 then RFM='F';
      end;
    if (PCTL_VALUE24)='03:214-312' then do;
         if buy18=0 and buy12=0 and buy6=0 then RFM='G';
         if buy18 ge 1 or buy12 ge 1 or buy6 ge 1 then RFM='H';
           if buy6=1 and buy12=1 and buy18=1 then RFM='I';
      end;
      if (PCT1_VALUE24)='04:312-high' then do;
         if buy18=0 and buy12=0 and buy6=0 then RFM='J';
         if buy18 ge 1 or buy12 ge 1 or buy6 ge 1 then RFM='K';
           if buy6=1 and buy12=1 and buy18=1 then RFM='L';
      end;
run;
```

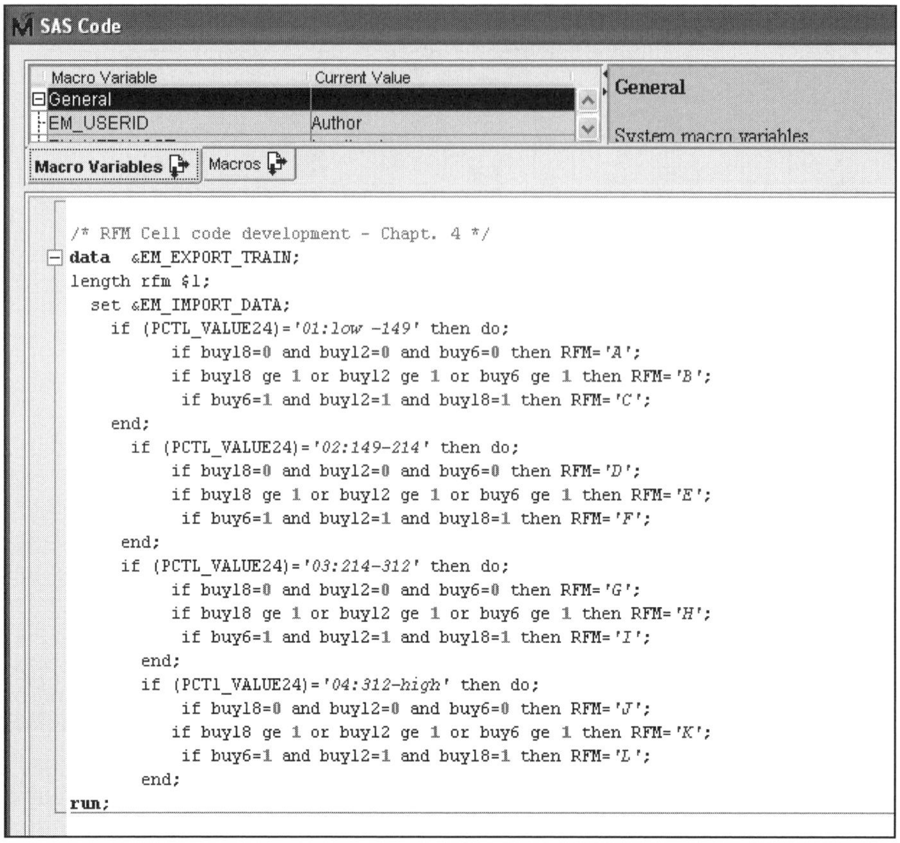

Figure 4.6 RFM Cell Develop Project—SAS Code Node Entry (*continued*)

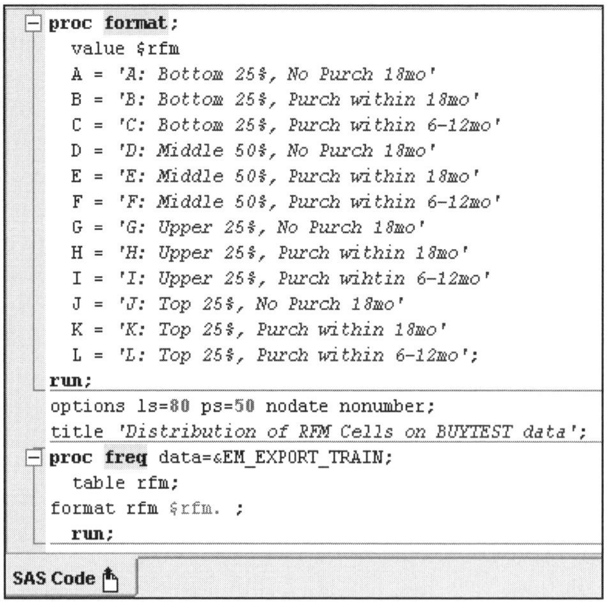

```
proc format;
  value $rfm
  A = 'A: Bottom 25%, No Purch 18mo'
  B = 'B: Bottom 25%, Purch within 18mo'
  C = 'C: Bottom 25%, Purch within 6-12mo'
  D = 'D: Middle 50%, No Purch 18mo'
  E = 'E: Middle 50%, Purch within 18mo'
  F = 'F: Middle 50%, Purch within 6-12mo'
  G = 'G: Upper 25%, No Purch 18mo'
  H = 'H: Upper 25%, Purch within 18mo'
  I = 'I: Upper 25%, Purch wihtin 6-12mo'
  J = 'J: Top 25%, No Purch 18mo'
  K = 'K: Top 25%, Purch within 18mo'
  L = 'L: Top 25%, Purch within 6-12mo';
run;
options ls=80 ps=50 nodate nonumber;
title 'Distribution of RFM Cells on BUYTEST data';
proc freq data=&EM_EXPORT_TRAIN;
  table rfm;
format rfm $rfm. ;
  run;

SAS Code
```

Notice the macro variables &EM_EXPORT_TRAIN and &EM_IMPORT_DATA. These macro variables point to an exported training data set and the data set imported by the SAS Code node, respectively. These macro value definitions can be viewed in the window just above the Code window. You can use explicit data set names if you desire; however, this approach is a bit more data-driven. The SAS FORMAT statements allow you to create labels on the levels of the RFM code values A through L and be more descriptive for presentation purposes. **Step 7:** The FREQUENCY procedure will produce a distribution of each RFM code value when this SAS code is run. Run the SAS Code node by either right-clicking and selecting the **Run** option or the **Run** button in the upper right corner of the SAS Enterprise Miner command screen. Once the SAS code has completed, a dialog box should appear indicating a successful run and asking if you want to view the results. Click the **Yes** button. The Output window displays and it should look like the one in Figure 4.7. When you place a PROC PRINT or anything that would generate an Output listing in SAS, it is placed in this Output window within the SAS Code node Results window. The frequency distribution of our newly created RFM cells should appear.

Figure 4.7 RFM Cell Develop Project—SAS Code Node Output Tab

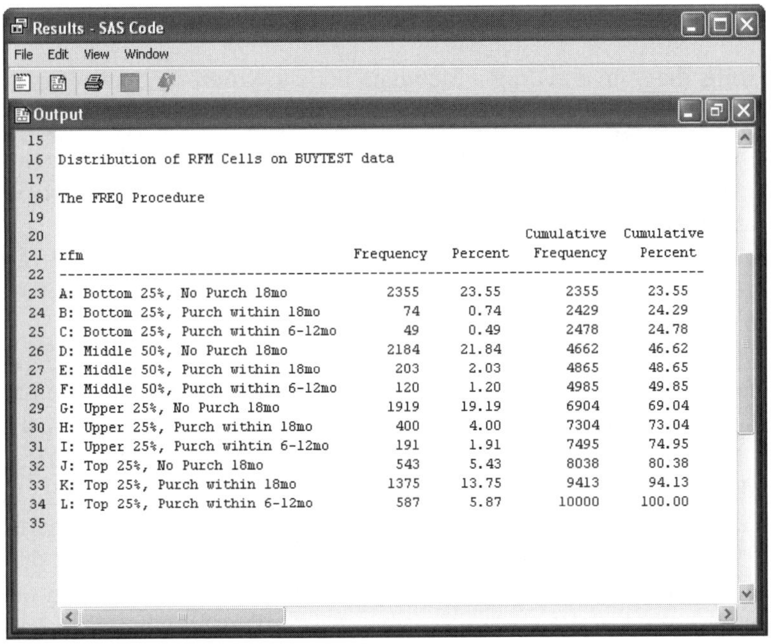

Results - SAS Code
File Edit View Window
Output

Distribution of RFM Cells on BUYTEST data

The FREQ Procedure

rfm	Frequency	Percent	Cumulative Frequency	Cumulative Percent
A: Bottom 25%, No Purch 18mo	2355	23.55	2355	23.55
B: Bottom 25%, Purch within 18mo	74	0.74	2429	24.29
C: Bottom 25%, Purch within 6-12mo	49	0.49	2478	24.78
D: Middle 50%, No Purch 18mo	2184	21.84	4662	46.62
E: Middle 50%, Purch within 18mo	203	2.03	4865	48.65
F: Middle 50%, Purch within 6-12mo	120	1.20	4985	49.85
G: Upper 25%, No Purch 18mo	1919	19.19	6904	69.04
H: Upper 25%, Purch within 18mo	400	4.00	7304	73.04
I: Upper 25%, Purch wihtin 6-12mo	191	1.91	7495	74.95
J: Top 25%, No Purch 18mo	543	5.43	8038	80.38
K: Top 25%, Purch within 18mo	1375	13.75	9413	94.13
L: Top 25%, Purch within 6-12mo	587	5.87	10000	100.00

Now when you rerun the SAS Code node, instead of labels A through L in your PROC FREQ, the formatted values will appear in the output instead. Be sure to alter your PROC FREQ to include the new format. Figure 4.7 shows the output with the formatted values of RFM. The macro variable &EM_EXPORT_TRAIN now contains the name of the data set that has the scored RFM values. When you create data sets using the SAS Code node you not only create the fields and data elements, you also create any original settings you placed in the Input Data Source node, like target variables, or other attributes you might have set. This allows those settings to follow your data so you don't have to reset them later in your mining project. This is what is commonly called metadata. We'll now embark on how to use this newly created RFM cells in a segmentation of customers.

4.4 Tree-based Segmentation Using RFM

Now that we have a single variable RFM that contains attributes of recency, frequency, and monetary value of these customers, let's use them in a typical segmentation scenario. Imagine a marketing manager desiring to increase the revenue potential of the customer base and to execute a few direct marketing campaigns to supplement this initiative. The manager would like to offer the best suited offer for each individual customer; however, in the customer data (e.g., the BUYTEST data set) we have 10,000 unique customers. Designing a unique offer for each of the 10,000 customers would be a daunting task. However, the manager indicates that the marketing department could effectively design a program that contains several offers but they would need to match with certain segments of customers. This is where you, the analyst or database marketer or whatever title you like, comes in and offers to help the manager segment the customer base into groups suited for the manager's campaign programs. The ideal situation for a marketing program is when a marketer can offer the right set of products or services to the right customer at the right time when the customer needs them. So, let's reopen the RFM Example project (if you had closed it) and see how we might construct several segments using the RFM cells and perhaps several other attributes about our customer base.

Step 8: Open the RFM Cells diagram and drag a Metadata node to your process flow diagram. Connect the SAS Code node to the Metadata node. Highlight the Metadata node and update the process flow path by clicking the ⟳ icon in the upper right corner of the SAS Enterprise Miner window. Now, in the Metadata node, click the **Imported Data** selection of the property sheet and a window appears indicating the imported data from the prior node. This data set now contains all of the fields in the original data set, including the newly created RFM field. The RFM variable should be set to Input as the model role and nominal for the measurement type. Click **Variables** in the property sheet and change the role of the RESPOND variable to **target**.

Now let's say the marketing manager would like to make some special offers on this customer set. The manager would like to include the fact that some customers have responded in a previous campaign. Let's choose the fields RFM, RESPOND, AGE, SEX, CLIMATE, INCOME, OWNHOME, FICO, and MARRIED for this segmentation study. **Step 9:** Now connect to a Data Partition node and connect the Input Data Source node to it. In the Data Partition property sheet, set the percentages to 70% training, 20% validation, and 10% for the test set. Now set the Partition Method to **Stratification**. Now open **Variables** in the Data Partition property sheet, select the **RESPOND** variable, and set it to **Stratification**. This will ensure that the proportion of the RESPOND variable is approximately the same in the three data set partitions of TEST, VALIDATION, and TRAINING.

This process flow is now set up to partition the original data set of 10,000 observations into a training, validation, and a test set each containing sampled proportions of the target variable RESPOND so that no missing cases exist in each of the three data sets. The training and validation data sets are used to train and fine-tune the model, whereas the test data set does not *see* the model at all and thus represents new cases. However, we actually know the target values for this so we can compare the modeled results to the actual values. We will segment the training and validation data sets using a Tree node. This is what is commonly called a Classification Tree, because we are attempting to classify all the observations according to each customer's RESPOND value, depending on the values of the input variables we have selected. When the target response variable is categorical, then the tree is a Classification Tree, and if the target response is numeric, it would be a Regression Tree. The exception is a binary numeric target response such as "0" vs. "1", which would still be considered a Classification Tree. What we are intending is to have a classification scheme in which each level of RESPOND (0 or 1) should have a set of rules according to the values of the variables we've selected. Then, we'll select those variables according to the rules in the decision tree and the RFM scores we created to target specific audiences.

Step 10: Place a Decision Tree node on the diagram and connect the Data Partition node to it. In the property sheet of the Tree node, set the following settings given in the Decision Tree property sheet shown in Figure 4.8.

Figure 4.8 Advanced Property Sheet Tree Settings for RFM Segmentation

Property	Value
Node ID	Tree
Imported Data	
Exported Data	
Variables	
Interactive	
Splitting Rule	
Criterion	ProbChisq
Significance Level	0.2
Missing Values	Use in search
Use Input Once	No
Maximum Branch	2
Maximum Depth	6
Minimum Categorical Si	5
Node	
Leaf Size	5
Number of Rules	5
Number of Surrogate R	0
Split Size	
Split Search	

Property	Value
Split Search	
Exhaustive	5000
Node Sample	5000
Subtree	
Method	Assessment
Number of Leaves	1
Assessment Measure	Lift
Assessment Fraction	0.25
P-Value Adjustment	
Bonferroni Adjustment	Yes
Time of Kass Adjustme	Before
Inputs	No
Number of Inputs	1
Split Adjustment	Yes
Output Variables	
Variable Selection	Yes
Leaf Variable	Yes
Leaf Role	Segment
Performance	Disk

Variables: Select only RFM, RESPOND, AGE, INCOME, OWNHOME, FICO, MARRIED, CLIMATE, and SEX to a status of **use** and all other variables to a status of **not use**.

Criterion: Select **ProbChisq** for the splitting criteria, and the following items as in Figure 4.8.

Advanced Property Settings: Select the Assessment Measure **Lift**. Now, drag an Assessment Node to your diagram and connect the Tree node to it. Your process flow diagram should now look like Figure 4.9.

Figure 4.9 RFM Segmentation Process Flow Diagram

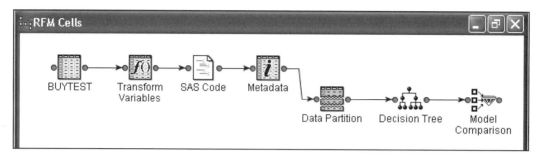

Step 11: Add a Model Comparison node and connect the Decision Tree node to it. You should now be able to run the entire flow by right-clicking the Model Comparison node or the Tree node and selecting the **Run** option. After you run the node, you should be able to see the results of the tree from the Decision Tree node's results. The Results viewer displays a set of windows, with one of them being a tree view where you can see the percent correct classification at each node and the stems with their rule, etc. **Step 12:** This view is shown in Figure 4.10. This view indicates at each node what the decision for splitting is and gives you a set of decision rules at each node. This is a profile of the RESPOND variable (0 and 1) from the input variables we selected. The variables at the top portion of the tree are most important at determining a responder of 1 vs. 0, whereas the variables near the bottom of the tree impact responders much less.

We can now see how this new segmentation model works when scored on the TEST data set. Remember in the process flow diagram we set up the Partition node and it broke up the original data set into three groups using the RESPOND variable as a stratification variable. This ensures that the proper proportion of 1s and 0s in the RESPOND variable are distributed in the TEST, VALIDATION, and TRAINING data sets as closely as possible to the original data set. The TEST data set was not used in building or fine-tuning the model. So, we can now *score* this TEST data set with the Tree segmentation model. **Step 13:** Now, drag a Score node and attach the output of the Decision Tree node to the Score node. Select the option in the Score property sheet to score the TEST data by selecting **Yes**. You can now run the Score node and it will score all three data sets including the TEST data set and apply the tree model. A comparison of the actual RESPOND vs. the predicted RESPOND can be made on this data set since this data was held out of the modeling effort and simulates newly scored records; however, we actually know which records have the proper RESPOND levels. **Step 14:** Right-click the Decision Tree node and select the **Results** option. The Results viewer of the tree model has several windows. You can select several chart options in the Score Ranking Overlay window. If you select the **Lift chart** option, a chart shows the lift at each decile of the response variable for both the TRAINING and VALIDATION data sets.

Figure 4.10 RFM Cell—Respond Segmentation Decision Tree Viewer

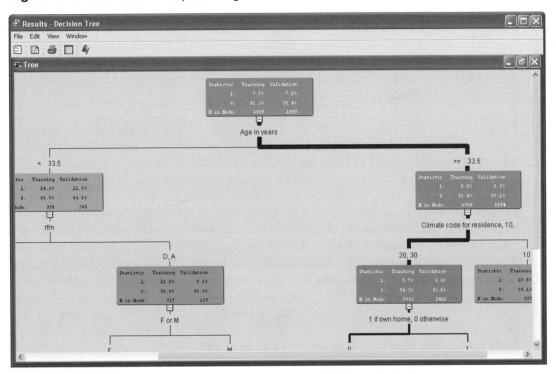

Step 15: If you open the results from the Score node, use the View menu and select **Graphs** and then the **Bar Chart** option.

Figure 4.11 SCORE Node Bar Chart Results

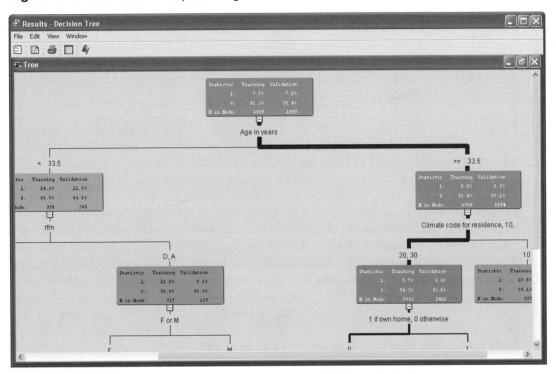

In the Model Comparison node, you can select the other graphic options and obtain charts of percent captured, the lift value, etc. The lift chart shows at each decile on each of the three partitioned data sets, the factor of improvement obtained from the tree model as compared to a baseline. Figure 4.11 shows a lift chart for the tree model; in the first decile of the TEST data, the factor improvement is between 1.5 and 2.5 times the value of the baseline. This indicates that individuals in the first decile are 1.5 to 2.5 times more likely to respond than if the individuals were chosen at random.

Back to our marketing manager, we can now select sets of RFM levels and other demographics from the model to start designing specific offers for each select group. For example, the first most important variable is AGE in years. So, the tree model splits the data on the variable AGE at a value of 33.5 years. If we select an age less than 33.5 years old and choose the level of RFM scores of A, D, E, F, G, J, K, and L, these are better responders for this age group. A custom marketing offer could be designed that is age appropriate for this group and also appropriate for other demographics that are not explicitly in the tree model. We can select the highest values of the P_RESPOND variable and when AGE is less than or equal to 33.5 and then review other demographics of this group. We haven't actually done any new data scoring but the beauty of this segmentation model is that we can use it for the existing data records, or perhaps new records where we don't know the RESPOND such as in the remainder of the database that the campaign was not run on and we can do a test campaign with these sets of RFM values and age groups.

Figure 4.11 Assessment Lift Charts from Tree Model

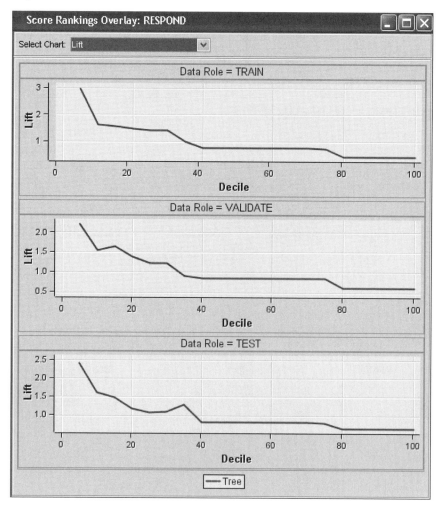

4.5 Using RFM and CRM—Customer Distinction

RFM scores usually are typically a good classification scheme for distinguishing and differentiating customers according to their three attributes. This does not mean, however, that the *future* customer buying behavior will behave just like their current RFM score; however, if these RFM scores are tested, then they can become predictive. It's far more convenient, however, to create a predictive model than to create RFM cells or scores and test them to find out if they are predictive or not. But, for customer distinction, RFM can be a very good choice. Remember, since RFM is based on past behavior, then the scores that are determined are classifying how customers *have* behaved. Let's take a look again at the distribution of RFM scores. Back in Figure 4.7, the distribution of RFM scores indicates the following characteristics:

- About 25% of the customer base has either not purchased within the last 18 months or is in the bottom 25% of purchases in the 6- to 18-month time frame.

- About 25% of the customer base has purchased middle-of-the-road in value within the past 6 to 18 months.

- Another 25% has purchased within the top 25% also within the past 6 to 18 months.

- And the remainder has a purchase in the top 25%.

What is interesting is that the middle and upper revenue segments have around 20%, 18%, and 5% (RFM cells D, G, and J) that have not purchased anything within the last 18 months! This is an excellent opportunity to get a fairly good purchaser to once again purchase something. The cells E, H, and K represent customers who have purchased a good amount in revenue but have only done so in the last 18 months. Therefore, this group is a potentially *growable* group of customers. By growable, I mean that they probably can or will purchase given the right items or perhaps a good sale item if offered. The bottom group, probably will need to have an offer they really can't refuse in order for them to purchase, and the very top keep coming back so perhaps a nice thank you is in store for them to keep them loyal, happy purchasers. These major themes of customer segments indicate that there are definite groups of customer distinctions that exist in this database and addressing those unique groups is the key to good customer relationship management. What marketers need are methods that separate customers into unique and distinct segments so that marketing plans can be identified that also uniquely satisfy the customers within those segments to the fullest extent possible.

Let's look at yet another way of viewing customer distinction. What might be ideal is if you knew the attitudes that each of your customers has and their preferences. Taking this data along side of your demographic data, you would have a much more complete and rich set of data to segment and mine than if you had only the demographic data alone. Now looking on the side of reality, one will rarely find a customer database in which each customer record contains both demographic and attitudinal information. So, how does one go about obtaining such information without embarking on a survey for every customer in the entire database? To survey all customers in a database would be very costly and time consuming and not every customer will respond. One method of solving this problem might be to survey some of the customers in the database and then build some sort of model of those customers whereby you have both sets of attitudinal and demographic data and then score the remainder of the database so that you would have estimates on the entire database. We perform this analysis using a SOM neural network in Chapter 11.

Some relatively new statistical techniques perform this type of analysis. These techniques are typically referred to as latent analyses (McCutcheon 1987, and Hagenaars and McCutcheon 2002). Latent class analysis (LCA) is a multivariate technique that can be applied to cluster, factor, or regression analyses. The latent part of the model construction is created from indicator variables and used to form clusters, factors, or predict dependencies in a regression. LCA divides the records in your database into latent classes, which are *conditionally independent*, meaning that the variables of interest are uncorrelated within any one class. Classes are considered latent because they are not directly observable but rather are identified based on a function of a set of indicator variables. Even though these analytical techniques came about in the middle-1980s, they were not used too much because of their intense computational requirements. Since the advent of much faster computers, the techniques of LCA are now more available for desktop or even laptop computers. SAS currently does not offer LCA procedures as such; however, an analysis that is similar in nature to LCA can be derived that mimics the general behavior of LCA. Very recently in the SUGI 31 proceedings, a paper was given showing how SAS DATA step code and the CATMOD procedure could be used in a SAS macro to perform LCA analysis (Thompson 1996). Later in Chapter 11, we'll look at a customer distinction example using a SOM data mining technique.

4.6 Additional Exercise

As an additional exercise, use the 500 customer survey questions set and create a predictive model to predict the responses of the customers, and score the remainder of the data set with the predictions. This may or may not work as 500 is very few rows of data with the number of question responses. You might consider grouping the questions even further into two or perhaps three levels and using that as a target variable.

4.7 References

Hagenaars Jacques A. and Allan McCutcheon, eds. 2002. *Applied Latent Class Analysis.* Cambridge, UK and New York, NY: Cambridge University Press.

Hughes, Arthur Middleton. 1996. *The Complete Database Marketer: Second Generation Strategies and Techniques for Tapping the Power of Your Customer Database.* Rev. ed. New York: McGraw-Hill.

Libey, Donald R. 1994. *Libey on Recency, Frequency, and Monetary Value.* Maryland: Libey Publishing Inc.

McCutcheon, Allan L. 1987. *Latent Class Analysis: Quantitative Applications in the Social Sciences*, Series 64. Thousand Oaks, CA: Sage Publications.

Pyle, Dorian. 2003. *Business Modeling and Data Mining.* San Francisco: Morgan Kaufmann Publishers, Inc.

Rud, Olivia. 2000. *Data Mining Cookbook: Modeling Data for Marketing, Risk, and Customer Relationship Management.* New York: John Wiley & Sons, Inc.

SAS Enterprise Miner, Release 5.2. Cary, NC: SAS Institute Inc.

Thompson, David M. 2006. "Performing Latent Class Analysis Using the CATMOD Procedure." *Proceedings of the Thirty-first Annual SAS Users Group International Conference.* Cary, NC: SAS Institute Inc. Paper no. 201-31.

4.8 Additional Reading

Berry, Michael J. A., and Gordon S. Linoff. 2004. *Data Mining Techniques: for Marketing, Sales, and Customer Relationship Management.* 2d ed. New York: John Wiley & Sons, Inc.

Segmentation of Several Attributes with Clustering

5.1 Motivation for Clustering of Customer Attributes: Beginning CRM

As we begin to look at the motivation of using clustering techniques for CRM, we've seen in Chapter 4, "Segmentation Using a Cell-based Approach," that one method of segmentation can be done using a Self-Organizing Map (SOM) Neural Network to accomplish a type of clustering. In this chapter we will look at the technique of clustering as some of the basic understanding of how clustering is measured was discussed in Chapter 3, "Distance: The Basic Measures of Similarity and Association." The concept of distance and similarity come into play as we attempt to find groups or segments within our database of customers, healthcare patients, fraud cases, and the like that all share some sort of similarity with each other.

Clustering is a technique that is typified as undirected data mining. It's undirected because there is no variable on the data set that we are trying to predict (e.g., no response variable). Sometimes, data sets are rather complex in nature and no apparent pattern seems to appear using other techniques like those that we discussed earlier: Decision Trees, or perhaps a Regression, or even a Neural Network. One method of quasi-directing a clustering technique is to allow the clustering algorithm to use only specific sets of variables that the data miner would like to use, and this will force the algorithm to use on those variables of interest in the clustering session. The natural tendency for humans when faced with a complex task is to attempt to break it down into much smaller bit-sized pieces, each of which hopefully are simpler in nature than the entire data set as a whole. In the context of CRM, the task of finding groups of customers inside your database that are similar in some way so that specific marketing and sales programs can be designed just for them is something that perhaps clustering could address.

The main question to answer first is how would you define similar customers? In the example in Chapter 1, "Introduction," children at an elementary school were sorted by their numeric sequence while standing in line; the measure of similarity was where the child appeared in a line sequence. All the number fours, threes, twos, etc. were grouped to form four teams, for example. So, the definition of what is a similar customer is probably closely related to the purpose for the groups or segments. This means that several or perhaps many cluster segments may exist for each data set of customers depending on what is selected for the measure of similarity. For example, if the desired objective in a CRM project is to understand and profile customers in a database, then perhaps the measure of similarity should be variables or combinations thereof that help describe how customers are different or similar to each other. For business customers, perhaps the type of industry, the size of the company in number of employees, or the amount of revenue, etc. can be measured.

Other metrics of interest could be how many times customers responded to a marketing campaign in the last six months, or which Web pages they visited on your Web site; how long they spent on-line, and whether they were just searching for information or visiting product or service areas on your Web site. All of these types of metrics help to classify a customer with respect to some type of behavior of interest. A single data set could possibly contain many varying cluster segments depending on what was used as inputs to measure similarity. This could be useful in the business context as different types of segmentation could be performed on the same set of customer records. Combinations of these segments could be used in conjunction with each other in order to accomplish a specific business purpose. The next section will look at an example test case with over 100,000 customers.

5.2 How Can I Better Understand My Customer Base of Over 100,000?

In this example, we'll start by taking the business context of the problem first, and then see how clustering, segmentation, and data mining aid in solving the business problem. One of the reasons that clustering and segmentation of a set of customers is so popular is that this technique allows the grouping of customers to take place along several dimensions simultaneously. For the business case in this problem, a marketing manager needs to send out several communication briefs to the customer base of a little over 100,000 business-to-business (B-B) customers about a new product introduction. This marketing manager would like to mail customized information notices to each customer and perhaps follow up with some of them with phone calls; however, the manager knows that 100,000 customizations would be a logistical headache, to say the least, let alone the cost associated with customizing each of the direct mail pieces. Our task then is to segment this customer base into groups of *like* customer sets and to customize for each of the customer segments. **Step 1:** To start this exercise, start up SAS Enterprise Miner Version 5.2 and copy the data set called CUSTOMERS to the default SAS library SAMPSIO typically located in the folder in `C:\Program Files\SAS\SAS 9.1\dmine\sample`. This data set is on the data CD in the Chapter 5 folder and the Segmentation Example subfolder. The description of this data set is found in Appendix 1.

Process Flow Table 1: B-B Segmentation

Step	Process Step Description	Brief Rationale
1	Start Enterprise Miner and copy the data set to the SAMPSIO library.	
2	Create a new project process flow diagram.	
3	Add the CUSTOMERS data set to the Data Sources folder.	
4	Perform a Data Assay on the EST_SPEND variable.	Reviewing how estimated spend is distributed for possible transformations.
5	Transform EST_SPEND to look more like a *normal* distribution.	Making the distribution of some variables look more like a normal distribution. This will aid the model building process strongly and produce more desired results.
6	Transform several other variables. The variable LOC_EMPLOYEE is transformed into quantile bins.	Making the distribution of some variables look more like a normal distribution. This will aid the model building process strongly and produce more desired results.
7	Filter out bad data points.	Removing bad data from the analysis.
8	Add a Clustering node to the data mining process flow.	
9	Specify clustering options.	Selecting initial options and reviewing results. This process is usually iterative.
10	Find what the clustering run found and profile the cluster segments.	Once profiled, a determination can be made if the cluster segments are useful for the analysis at hand.
11	Take note of the number of customer records in each segment.	Testing if the segments are relatively well proportioned (some not too large or too small).
12	Observe the Distance plot and understand the variables that might make up dimensions 1 and 2.	Seeing how the clusters relate in their transformed space.
13	Use the Segment Profile node to aid in segment profiling.	Segment profile allows other types of profile attributes not found in the cluster node.
14	Open the Variable Importance Plot.	Observe each variable's contribution to the cluster model.

Step 2: Create a new project entitled Segmentation Example and a new process flow diagram called B-B Segmentation. To start the exercise on the right foot, a data assay is in order. Do you remember back in Chapter 2, "Why Segment? The Motivation for Segment-based Descriptive Models," the concept of the data assay was discussed? Well, it's back again, and this time we need to understand the variables on this data set as well as any missing values, etc. One of the

first things you can do is to look at some basic properties of the variables, and this can be easily done using the Input Data Source node. **Step 3:** Expand the Data Sources icon and right-click **Create a Data Source**. The Data Source Wizard will appear on your computer screen so it looks like the one in Figure 5.1. Follow the directions to add the CUSTOMERS data set located in the SAS library SAMPSIO. In the Apply Advisor Options section, click the **Advanced** button and then proceed.

Figure 5.1 Input Data Source Wizard Window

The Data Source Wizard – Metadata Source window now appears and you can explore several variables. **Step 4:** Highlight the EST_SPEND variable, and click the **Explore** button to show the approximate distribution of the highlighted variable(s), as shown in Figure 5.2. Other variables can be viewed prior to entering the data in the Data Source Wizard. Now complete the wizard and add an Input Data Source node to your process flow diagram. Be sure to edit the variables and change PURCHLST and PURCHFST to Ordinal, and the variable PUBLIC_SECTOR to Binary. Click the **Data Source** label in the Property/Value dialog box, and select the CUSTOMERS data set in the SAMPSIO library.

Figure 5.2 Approximate Distribution of Estimated Spend Dollars in Distribution Explorer Node

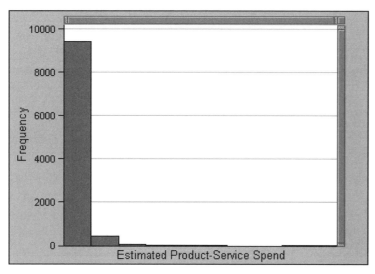

This distribution in Figure 5.2 looks highly abnormal; e.g., it does not look like a bell-shaped curve, which would give indication that the distribution is from a normal distribution. If you recall the discussion of the *k*-means algorithm back in Chapter 3, "Distance: The Basic Measures of Similarity and Association," one of the characteristics of the *k*-means algorithm is that if a distribution is very abnormal, then the tails of the distribution tend to be their own set of clusters and typically defeat the purpose of clustering for CRM. To fix this situation, let's transform the EST_SPEND distribution so that it looks like a bell-shaped curve and is closer to a normal distribution. **Step 5:** Drag a Transform Variables node onto the diagram workspace and connect the Input Data Source node to it. Highlight the Transform Variables node and then click the Formula builder property sheet. Transform the variable EST_SPEND by clicking the Create icon. The variable should be called TRANS0, and you can change the label as well. To view the newly transformed variable, click the **Generate Plot** button. This window now allows you to view the data and now select the Plot icon and plot a histogram of the transformed log (EST_SPEND). The transformed value of EST_SPEND is shown in Figure 5.3. Notice that this distribution looks like it has two distributions instead of one; this is what is typically called a bi-modal distribution. In the process of trying to segment this customer base, perhaps we could also uncover why this distribution is bi-modal in nature and what attributes might explain this nature if possible.

Step 6: Now, if you have deselected the Transform node, select it and let's transform some of the other variables that are similar in nature to EST_SPEND (for example, last year's and this year's revenue and total revenue fields). Transform them in a similar fashion, selecting the Log option. For the loc_employee field, select Quantile and select 4 quantile bins; Figure 5.4 shows the selected variables to transform. The other variable that you should transform should look like Figure 5.3. What we are doing here is transforming the distributions into a form that will enable the algorithms such as regressions, neural networks, and decision trees to perform better than if the variables were not transformed.

Figure 5.3 EST_SPEND Transformed to Approximate Normal Distribution

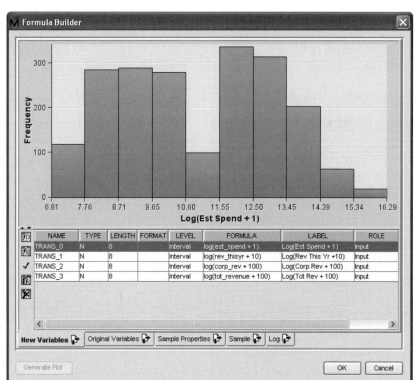

Figure 5.4 All Transformed Variables in the Node

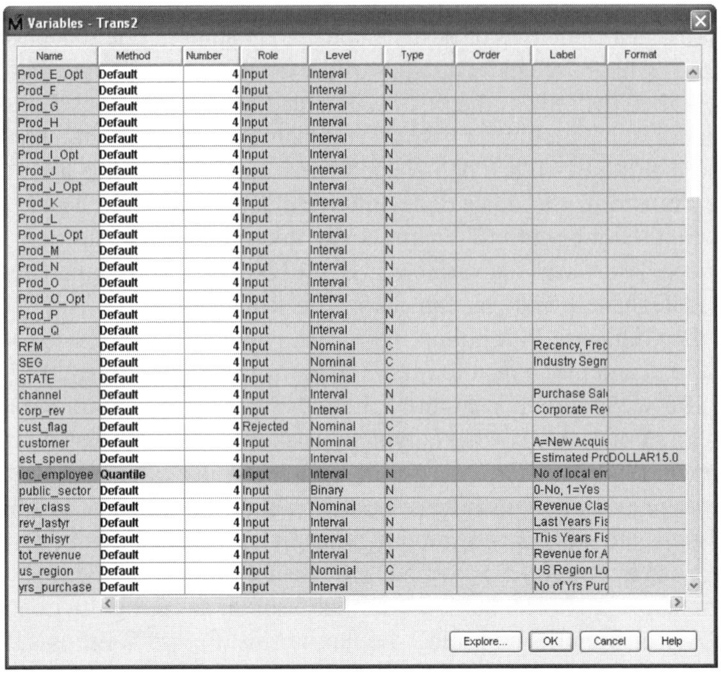

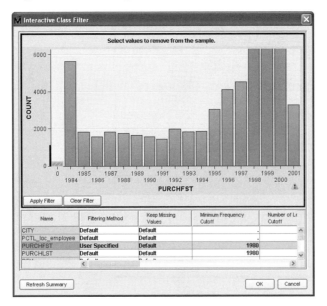

Step 7: Now, there are a few variables that we need to clean up a bit prior to performing an analysis, so drag a Filter node onto the workspace and connect the Transform Variables node to it. Now select the Variables label. The first year and last year Purchase fields have some data in them that appears to be either entered incorrectly or corrupt. The low end of this distribution has some values around 125 and we know the year 125 is incorrect, so we'll want to omit these incorrect values. To do this, click on the class variables property and click on the PURCHFST field. A window showing the variable's distribution will appear. You can stretch the axis and then highlight the low value, as shown in Figure 5.5. Do the same for the PURCHLST field. This will filter all observations in the data below the value you entered in these fields.

Figure 5.5 Filtering out Outlier Values

We can now add a Clustering node to the diagram process flow and connect the Filter node to it. We could approach the problem of segmentation by just taking a combination of RFM scores (see Section 5.4, "References" for a description) and some other variable like US_REGION and proceed to segment the customer base accordingly. However, this example is intended to show how a method like a cluster algorithm can perform this task with a set of variables simultaneously.

Step 8: Drag a Clustering node to the process flow and connect the output of the Filter node to it. In the Properties column, click the **Variables** label to open the window that lists all the available variables going into the clustering node. If you don't see a longer list of properties, use the View menu and select the Advanced Property Sheet check box.

The first item of interest is the Variables property. This is where you will select or reject variables to be used by the Clustering node to perform the cluster algorithm segmentation, and you can determine which variables are reported in the cluster graphs. Highlight all of the product fields (which we will be looking at in Chapters 9 and 10) as well as the original data fields that we transformed, so that the only ones we are keeping are shown in Figure 5.6. **Step 9:** The default in SAS Enterprise Miner is that the original variable from the transformed variables is normally dropped. You may want to see them so they can be used later on. Be sure that the Customer ID field is set to Use because this is how the clustering algorithm distinguishes customers. Now for the first pass, select **Standardization property→Range**; the standardization feature causes all of the selected USE variables to place the values of each variable between the 0 and 1. Also select the clustering method Ward and set all the property values and flow diagrams as shown in Figure 5.7. Run the Cluster node and view the output results by clicking the Results icon in the upper right SAS Enterprise Miner window.

Figure 5.6 Cluster Node Variables Tab Initial Variable Settings

Name	Use	Report	Role	Level	Type	Order	Label	Fo
RFM	Default	No	Input	Nominal	C		Recency, Frec	
PURCHFST	Default	No	Input	Ordinal	N		Year of 1st Pu	
PCTL_loc_employee	Default	No	Input	Nominal	C		Transformed:	
SEG	Default	No	Input	Nominal	C		Industry Segm	
STATE	Default	No	Input	Nominal	C			
PURCHLST	Default	No	Input	Ordinal	N		Last Yr of Purc	
rev_class	Default	No	Input	Nominal	C		Revenue Clas	
channel	Default	No	Input	Interval	N		Purchase Sal	
yrs_purchase	Default	No	Input	Interval	N		No of Yrs Purc	
TRANS_2	Default	No	Input	Interval	N		Log(Corp Rev	
customer	Default	No	Input	Nominal	C		A=New Acquis	
TRANS_0	Default	No	Input	Interval	N		Log(Est Spen	
public_sector	Default	No	Input	Binary	N		0-No, 1=Yes	
TRANS_1	Default	No	Input	Interval	N		Log(Rev This	
TRANS_3	Default	No	Input	Interval	N		Log(Tot Rev +	
us_region	Default	No	Input	Nominal	C		US Region Lo	
tot_revenue	No	No	Input	Interval	N		Revenue for A	
cust_flag	No	No	Rejected	Nominal	C			
rev_thisyr	No	No	Input	Interval	N		This Years Fis	
corp_rev	No	No	Input	Interval	N		Corporate Rev	
rev_lastyr	No	No	Input	Interval	N		Last Years Fis	
est_spend	No	No	Input	Interval	N		Estimated Prc	DOLL
Prod_Q	No	No	Input	Interval	N			
Prod_E_Opt	No	No	Input	Interval	N			
Prod_N	No	No	Input	Interval	N			
Prod_A_Opt	No	No	Input	Interval	N			
Prod_O_Opt	No	No	Input	Interval	N			
Prod_G	No	No	Input	Interval	N			
Prod_H	No	No	Input	Interval	N			
Prod_L_Opt	No	No	Input	Interval	N			
Prod_D	No	No	Input	Interval	N			
Prod_K	No	No	Input	Interval	N			
Prod_A	No	No	Input	Interval	N			

Explore... | OK | Cancel | Help

Figure 5.7 Cluster Node Initial Properties Settings and Flow Diagram

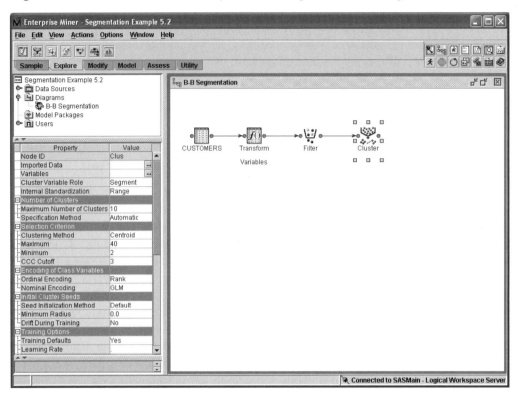

The Results window of the Clustering node looks like Figure 5.8 with a number of options available for profiling the segments found by the clustering algorithm.

Step 10: Our task now is to see if the clustering run performed satisfactorily and to profile these segments to discover what the clustering algorithm found. We also want to see if those customer segments will fit the basic business needs for the problem; that is, the marketing manager would like to send out customized messages and could possibly create perhaps 6 through 10 (perhaps no more than a dozen) different messages or offers. Therefore, about 6 through 12 is the practical business limit for the number of cluster segments that should be presented to that manager or marketing team. Click the Cluster node and then click the Results icon or right-click the Cluster node, and select the Results option. Notice that on the bottom left the pie chart shows the clusters as slices of the pie and shows the percentage of the training data set that falls within each cluster.

Step 11: With the exception of cluster 1, most clusters have around 20,000–25,000 observations or so (cluster 1 being the smallest with around 13,000 customer records). These clusters are roughly similar in size, which is generally a good thing for the marketing team to have somewhat similarly sized groups to work with. One of the next items you might want to employ is to view how the clusters relate distance-wise in the *transformed space*. Transformed space is the set of dimensions that map the original set of variables used as inputs to the clustering algorithm to the 0 through 1 set of transformed ranges that allows distance metrics on each variable to be comparable on the same scale as was discussed in Chapter 3. With the Cluster Node Results window open, click the View menu, and select Distance and Plot as shown in Figure 5.9. What you should now see is the Distance Plot window, and when you expand it the set of five clusters should appear like those in Figure 5.10.

Figure 5.8 Cluster Node Default Results Window

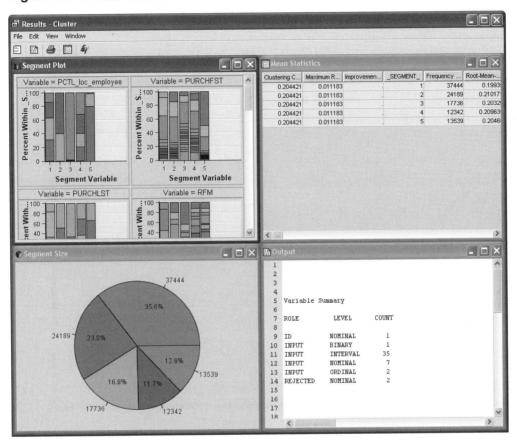

Figure 5.9 Cluster Node Default Results Distance Plot Selection

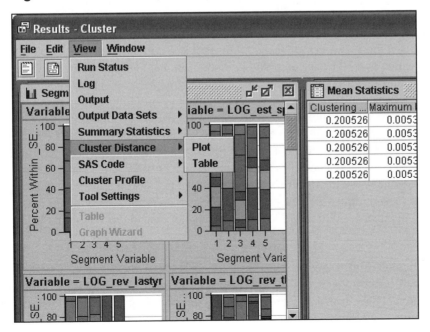

Figure 5.10 Cluster Node Distance Plot for the Five Clusters

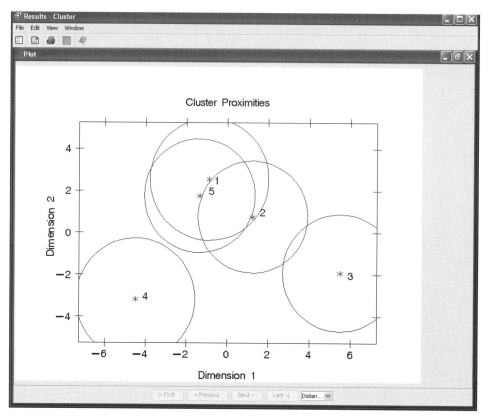

Step 12: This plot was generated by taking the distance matrix computed for the top two dimensions of the eigenvectors and using multidimensional scaling to obtain the plot. To aid in the process of understanding what differences these clusters have with each other is to profile cluster 3 versus cluster 4 or 5, for example. In that fashion, you may discover the set of variables that will dominate the x-axis of Dimension 1 in Figure 5.10. In a similar fashion, you could profile cluster 5 against cluster 1 to help discover the main factors of Dimension 2, the y-axis in Figure 5.10. Notice that these clusters are fairly well separated and although they look like they overlap, in actuality, however, they are completely distinct in that every customer record or observation will fall into exactly one and only one cluster segment. You should strive to have little overlap and well-separated clusters. This will be each cluster with a unique distinction not found in the other clusters. Other algorithms have what is typically called fuzzy clusters in which the probability of cluster membership is given, and these kinds of algorithms will be discussed later in Chapter 11, "Computing Segments Using SOM/Kohonen for Clustering." You want good separation of the clusters and the values of Dimensions 1 and 2 not to be too much greater than 100, or 150 positive or negative in absolute value. If you had not selected the variables and perhaps not done any standardization (like range or standard deviation), then the dimensions of Figure 5.10 could range from –5000 to + 5000, and the cluster distribution might have looked rather different.

Step 13: Another method of profiling the cluster segments is to use the Segment Profile node in conjunction with the Cluster node profiling capabilities. Drag onto your diagram a Segment Profile node and connect the Cluster node to it. Run the Segment Profile node and when complete highlight the node, and click the Results icon. You now have a set of profile plots of Segment Size (pie chart), Profile (set of plots that display each row as a segment comparing the distribution of various variables as you scroll the window to your right), and Variable Worth plots for each segment. In the Output window will be some basic statistics of each segment. An example of the Segment Profile Results window for these sets of clusters is given in Figure 5.11. These profile capabilities allow you to understand the role each set of variables has within each segment. For example, in Figure 5.11 notice that segment 3 and segment 2 have differing sets of variables that explain each segment from highest to lowest importance moving from left to right.

Figure 5.11 Segment Profile Results Window

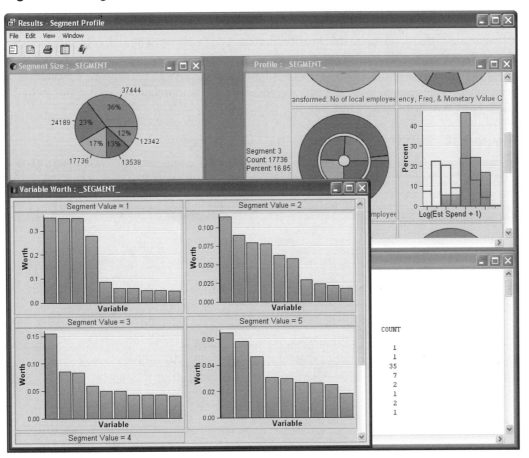

Now, if you minimize the Variable Worth, Segment Size, and Output windows, then expand the Profile window you will see for interval and class variables the same set of variables from left to right as you did in the Variable Worth window for each segment. For interval variables, you'll see a double histogram; the blue shaded area represents the within-segment distribution, while the outline histogram (typically red) represents the general population distribution so you can see how that variable differs from the overall population for that variable. For class variables, you will see a tree ring-like diagram that has two concentric rings; the inner ring represents the distribution of the total population, while the outer ring represents the distribution of the indicated segment. So, in Figure 5.12 the expanded view of the Segment Profile window is shown. You can see segments 2 and 3 and the first few variables that make up those segments. To see the other variables, use the horizontal scrollbar, and to view more segments, use the vertical scrollbar.

Step 14: Another useful chart plots the relative variable importance overall. This indicates which variables have the most impact at determining the clusters. To do this, click the Cluster node and open the Results window. Then use the View menu to select Cluster Profile, and then Variable Importance. A table showing the relative importance is shown. You can plot this by clicking the Plot Icon wizard and plotting the mean relative importance by each variable as a category. Figure 5.13 shows such a plot and indicates that transformed estimated spend, last year purchased, number of years purchased, and transformed total revenue are among the highest importance.

In the same fashion, continue this profiling of each segment until a relatively satisfactory description emerges that differentiates each segment. Then, each segment is sometimes labeled, other than just with numbers, by giving it a name that mimics or typifies the description, like technology pace setters, loyal and true, or perhaps lagging purchasers, and other catchy little names that will help act as a mnemonic.

Figure 5.12 Segment Profile Expanded Window View

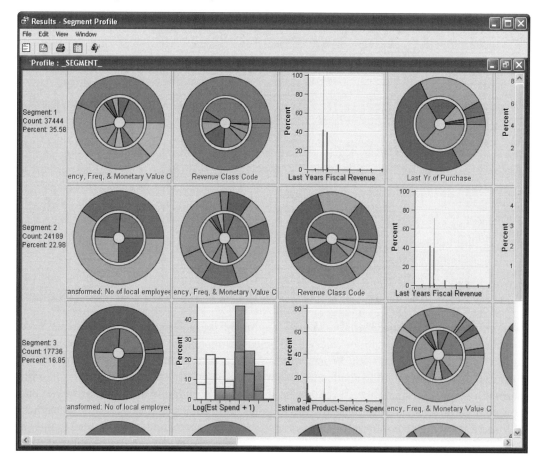

Figure 5.13 Plot of Variable Importance in Cluster Results

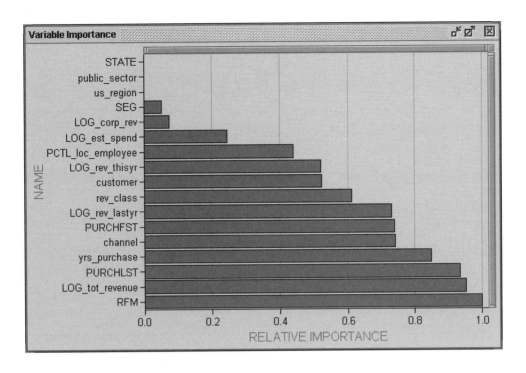

5.3 Using a Decision Tree to Create Cluster Segments

We now come to the place where the definition of clustering is not limited to the world of cluster algorithms; decision tree algorithms can also perform clustering as well, although they do it in a very different fashion to cluster algorithms. An empirical decision tree represents a segmentation of data that is created by applying a series of simple rules. Each split rule will assign a record or observation in a data set to a segment based on the value of an input such as salary, for example. These sets of rules split data into partitions, like salary less than $50,000 into one group and all other salaries into another group. At each split, there are several types of measures that could be employed, depending on the particular algorithm used, to measure the results of that data split. While it is out of the scope of this book to review all such algorithms, however, one in particular that will help accomplish our task of partitioning the CUSTOMERS data set is an algorithm called *Gini impurity*. Impurity, or rather purity, is a measure of how *pure* a population of observations is after a split in the data. A decision tree splits a set of data on a particular variable and then the split portions are measured for how pure the classification samples are. For example, consider Figure 5.14. A split of variable A subdivides the observations (or data records) into two groups. There are two categories, the dark and light shaded objects. This is considered to be a good split as the resulting populations in each of the split portions are *pure* for the most part (Berry and Linoff 2004, pp. 177–178). This measure of *pureness* is a close cousin to the measure of similarity we reviewed in Chapter 3.

Figure 5.14 Illustration of a Binary Split and Concept of Purity

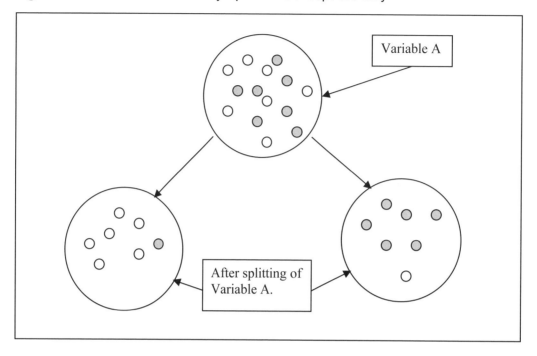

The parent node (top larger circle of Figure 5.14) contains an equal number of shaded and light circles and after the splits on variable A there is one misclassification in each left node. Therefore, a split of this type on variable A increases the purity in segmenting the light and shaded circles. Now, back to the Gini impurity; an Italian statistician, Corrado Gini, invented this measure of population diversity, which is used in biology and ecology studies. A pure population, according the metric, is the probability that two items chosen at random from the same population are in the same class (Berry and Linoff 2004, pp. 177–178). A completely pure population will have a probability of unity. SAS Enterprise Miner computes this metric, along with others, and is computed in Equation 5.1 (SAS Institute Inc. 2003):

$$I(\text{node}) = \left(1 - \sum_{i}^{\#ofclasses} \left(\frac{n_i}{N} \right)^2 \right)$$

(5.1)

where n is the number of class i cases and N are all cases in the node.

This computation can be optimized (or minimized if you subtract 1 for Equation 5.1) for greater purity or impurity, respectively. In effect, splitting the data in this fashion allows observations at each leaf node to be more similar than in other leaves of the Decision Tree. Again, a form of clustering has been done according to the measure of *pureness*. However, it should be noted here that in the algorithm of clustering discussed in Chapter 3, there is no target or response variable. This is the difference between *directed* versus *undirected* data mining. Clustering is considered *undirected* and a decision tree is *directed* because in clustering, there is no response, whereas in a decision tree, there is a response variable.

In a decision tree, a response variable is used as the variable in which to optimize purity. One can then select several response variables, create a decision tree from each, and then possibly combine the results. Perhaps another method would be to choose a single variable, which is known from previous profiles and analyses that yields a variable or two that is a strong classifier. Since RFM scores were created to do such a thing, and RFM was very high on the previous clustering exercise, RFM might be a good choice. So, let's create a decision tree to predict RFM with the intent of clustering.

We need to consider at this point if the number of levels of RFM is the right level of aggregation to compute a class target variable. So, without further ado, create a new diagram **(Step 1)** called Decision Tree Clustering and drag the CUSTOMERS data set from the Data Sources folder **(Step 2)**. The new process flow table is given below.

Process Flow Table 2: Decision Tree Clustering

Step	Process Step Description	Brief Rationale
1	Start SAS Enterprise Miner.	
2	Add the CUSTOMERS data set to your new process flow diagram.	Making a new diagram within a project.
3	Attach a StatExplore node to the CUSTOMERS data set.	Understanding distribution of the RFM variable.
4	Collapse the RFM variable into a smaller set of RFM levels.	Setting up target variable with appropriate number of levels for a predictive model.
5	Add a SAS Code node to the process flow diagram.	
6	Use SAS code to collapse RFM levels from 11 to 5.	
7	Add a Data Partition node to the diagram.	Stratified random sampling of the target variable RFM_NEW for training, test, and validation data sets.
8	Drag a Decision Tree node onto the flow diagram.	Developing a decision tree model to predict RFM_NEW variable.
9	Select all variables except product variables for the decision tree model.	
10	Review the decision tree model results.	

In this example, we'll want to make the RFM variable a target variable instead of just an input variable. **Step 3:** Attach a StatExplore node to the input data source of CUSTOMERS. Run the StatExplore node. When you open the Results window, the Output window will show the frequency distribution of the RFM target variable, among many other statistics. Figure 5.15 shows the results Output window with the RFM variable's results.

Figure 5.15 StatExplore Results Window of Target RFM Variable

```
Results - StatExplore
File  Edit  View  Window

  Output
  73
  74                     Formatted
  75  Variable    Role     Value    Frequency    Percent
  76
  77  RFM      TARGET       A         19402      18.3966
  78  RFM      TARGET       F         18214      17.2702
  79  RFM      TARGET       K         15672      14.8599
  80  RFM      TARGET       G         13457      12.7597
  81  RFM      TARGET       B          8330       7.8984
  82  RFM      TARGET       J          7700       7.3010
  83  RFM      TARGET       I          6338       6.0096
  84  RFM      TARGET       H          5619       5.3278
  85  RFM      TARGET       C          4401       4.1729
  86  RFM      TARGET       D          4266       4.0449
  87  RFM      TARGET       E          1995       1.8916
  88  RFM      TARGET                    71       0.0673
  89

  Chi-Square Pl...
```

Because there are A through K levels of the RFM variable, this number of levels could pose a problem trying to predict that number of levels as a target response. **Step 4:** So let's collapse some of the levels to reduce the number to a more manageable set. **Step 5:** Drag onto the SAS Enterprise Miner workspace a SAS Code node and connect the Input Data source node to it. You can open the SAS Code window by clicking the SAS Code property item and the … button as shown in Figure 5.16.

Figure 5.16 SAS Code Node Property Dialog Box Window

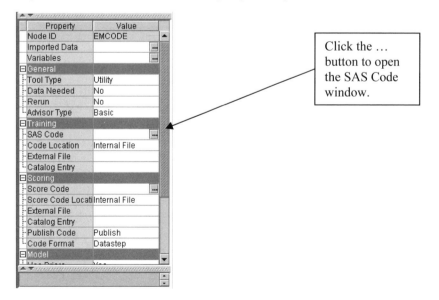

Click the … button to open the SAS Code window.

Step 6: Once you have opened the SAS Code node, place into the coding window the following code to reduce the number of RFM levels from 11 to 5 as shown in Figure 5.17.

Figure 5.17 SAS Code Node to Reduce RFM Levels

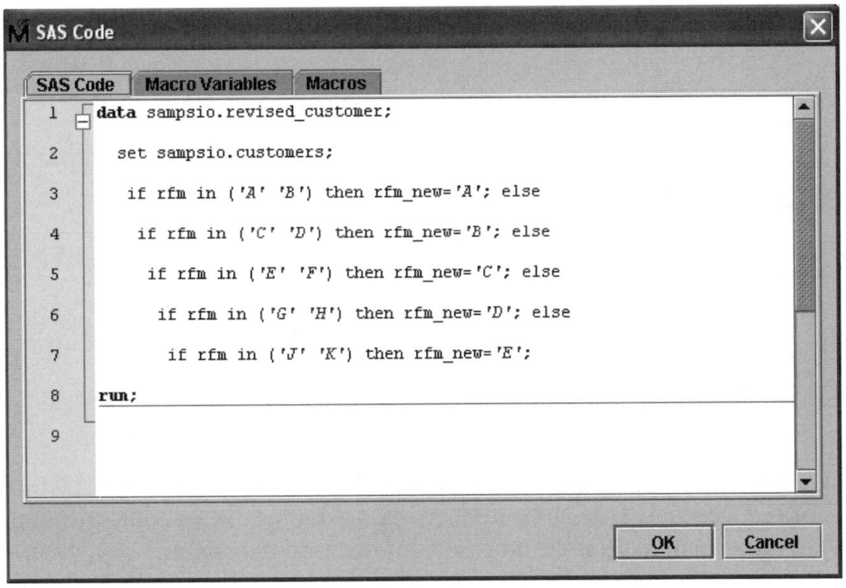

```
data sampsio.revised_customer;

  set sampsio.customers;

  if rfm in ('A' 'B') then rfm_new='A'; else

   if rfm in ('C' 'D') then rfm_new='B'; else

    if rfm in ('E' 'F') then rfm_new='C'; else

     if rfm in ('G' 'H') then rfm_new='D'; else

      if rfm in ('J' 'K') then rfm_new='E';

run;
```

After you have entered the code, run the SAS Code node icon in the diagram. This will generate the new data set called REVISED_CUSTOMER in the SAMPSIO library. You can now add this data source into the Date Sources folder and then drag it onto the diagram workspace so it can be used in the data mining process flow. Be sure that the new variable RFM_NEW is set to a target role in the Input Data Source node. What we will do now is to build a model that predicts the values of our RFM_NEW field (A through E). Then the rules in the decision tree that define each segment can be used as profile rules just as we saw in the previous example using the Segment Profiler node. **Step 7:** The next step is to put in our diagram a Data Partition node. A Data Partition node will statistically sample the original data source into three data sets; one for training a model, one for validating a model, and one for testing the scoring results, but is not used in building the model—e.g., a holdout sample. These data sets get carried along with the remainder of your process flow and will also be used during model assessment. In the Data Partition node, set the Partitioning Method to Stratified and the Data Set Percentages to 60, 30, and 10 percent for Training, Validation, and Test respectively. The Data Partition node property sheet is shown in Figure 5.18.

Figure 5.18 Data Partition Node Property Sheet Window Settings

Property	Value
Node ID	Part
Imported Data	
Variables	
Partitioning Method	Stratified
Random Seed	12345
⊟ Data Set Percentages	
Training	60.0
Validation	30.0
Test	10.0
⊟ Status	
Last Error	
Last Status	Complete
Needs Updating	No
Needs to Run	No
Time of Last Run	1/27/05 9:26 PM
Run Duration	0 Hr. 0 Min. 20.1

Step 8: Now, drag a Decision Tree node and connect the Data Partition node to it. The settings we want to select for our Decision Tree node are the Gini option for the splitting criterion and misclassification for assessment measure. Compare your decision tree settings to those in Figure 5.19 in the property sheet. Figure 5.19 is sectioned into two halves; in SAS Enterprise Miner, your property sheet is one scrollable section.

Figure 5.19 Decision Tree Node Property Sheet Window Settings

Property	Value
Node ID	Tree
Imported Data	
Variables	
Interactive Training	
Splitting Criterion	Gini
Significance Level	0.2
Missing Values	Use in search
Use Input Once	No
Exhaustive	5000
Leaf Size	5
Maximum Branch	2
Maximum Depth	6
Minimum Categorical §	5
Node Sample	5000
Number of Rules	5
Number of Surrogate F	0
Split Size	
SubTree	Assessment
Number of Leaves	1
Performance	Disk
Variable Selection	Yes

Property	Value
Performance	Disk
Variable Selection	Yes
⊟ Assessment Options	
Measure	Misclassificati
Percentage	0.25
⊟ P-Value Adjustment	
Bonferroni Adjustment	Yes
Time of Kass Adjustm	Before
Inputs	No
Number of Inputs	1
Split Adjustment	Yes
⊟ Variable Generation	
Leaf Variable	Yes
Leaf Role	Segment
⊟ Status	
Last Error	
Last Status	Complete
Needs Updating	No
Needs to Run	No
Time of Last Run	2/2/05 9:06 PM
Run Duration	0 Hr. 0 Min. 34

Step 9: Select all variables as input (NEW_RFM as target) and set all the *product* variables to not to use. Now, run the Decision Tree node. This should take a minute or so to complete. Once completed, you can view the Results window for the Decision Tree. The Results window will allow you to view resulting data sets, charts, and fitting statistics, as well as the completed Tree diagram. **Step 10:** Figure 5.20 shows the default Results window of the completed Decision Tree node.

Figure 5.20 Decision Tree Node Results Window

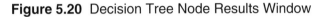

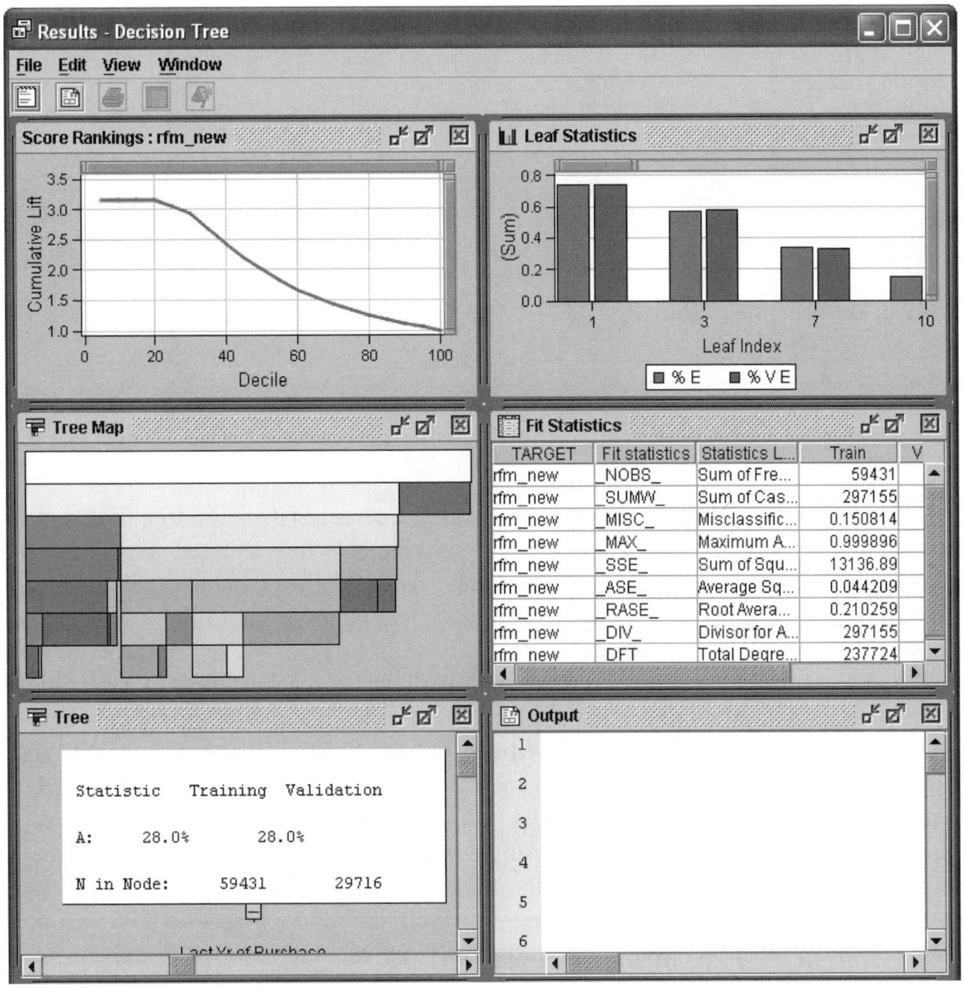

You can also view the data sets of the predicted results; the predicted NEW_RFM variable has a label of Into: NEW_RFM with the predicted levels of A through E. You also get the predicted probability of each level. For our cluster segments, these probability values are the probability of cluster segment membership. Figure 5.21 shows what the completed process flow diagram should look like. You can attach a Model Comparison node and run, and then view the Results window just like the Decision Tree node. This gives you additional information on the model performance. You can see how the training, validation, and test data sets compare with the levels of proper versus misclassifications in the NEW_RFM target response.

If you performed this kind of segmentation on a data set that is representative of a much larger data set, then you could generate scoring code to score the larger data set with these predicted values of NEW_RFM.

Figure 5.21 Completed Decision Tree Clustering Process Flow Diagram

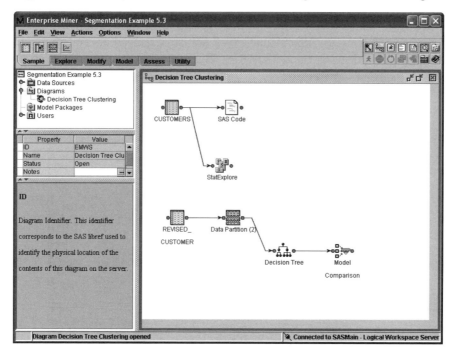

The analysis previously performed used what is called a *directed* decision tree model in that a *directed* data mining model is one where a target variable is to be predicted. The use of our model is not to predict but to cluster the records of our data set. When we clustered the data set using a clustering algorithm, that was an example of an *undirected* data mining model because no target variable is being used. Both of these techniques in SAS Enterprise Miner can produce scoring code that can be used as it stands in the process flow diagram or transferred to a data mart to score a larger set of data or a holdout data set that has not been run with the model. This data set, called the TEST data set in SAS Enterprise Miner, has not been used to build or fine-tune the decision tree model but can be used to see how well the model predicts the NEW_RFM on this set of data that was not used during the model building process. The TEST data set was partitioned when we used a 10% stratified sample in the Data Partition node. When you use a Model Comparison node as mentioned earlier, then actual versus predicted results on the TEST data set can be used to aid you in determining how well this model might behave on a much larger data set where the TEST data set is statistically representative of the larger set. The Model Comparison also allows you to see misclassification rates, lift charts, and the like.

5.4 References

Berry, Michael J. A., and Gordon S. Linoff. 2004. *Data Mining Techniques: for Marketing, Sales, and Customer Relationship Management.* 2d ed. New York: John Wiley & Sons, Inc.

SAS Enterprise Miner, Release 5.2. Cary, NC: SAS Institute Inc.

5.5 Additional Reading

For an article that describes how decision tree models can be undirected (unsupervised) and produce a hierarchical clustering, the article below is a good but very technical article.

Basak, Jayanta, and Raghu Krishnapuram. 2005. "Interpretable Hierarchical Clustering by Constructing an Unsupervised Decision Tree." *IEEE Transactions On Knowledge and Data Engineering* 17.1: 121–132.

For a general overview of decision trees, the following book is a classic.

Breiman, Leo, J. H. Friedman, R. A. Olshen, and C. J. Stone. 1993. *Classification and Regression Trees.* New York: Chapman & Hall.

Clustering of Many Attributes

6.1 Closer to Reality of Customer Segmentation

Many customer data sets have quite a few variables or fields, which can be used to perform clustering. Data analysts may find themselves in this situation that is closer to reality, as they approach customer segmentation. This is typically true for consumer databases as there are many, even hundreds or a thousand variables, which could potentially be used as inputs to cluster segmentation. Add a few more derived variables to this already rather large set and the task of variable selection could be daunting. Some potential questions concerning the approach of clustering might be: Should we consider using every attribute available to cluster our customer base? Is there a set of variables we *must* consider because there is a business need for those variables and they need to be included in the analysis? With the remaining set of variables other than ones that have to be there for a business reason, is there a methodology that can help determine which set of variables are most important and can best explain the data, given certain criteria? By the end of this chapter, questions such as these and others will have been answered, at least with a basic understanding, on the possible approaches that can be employed for potential solutions.

Our next example, from Data-Miners.com, is the NYTOWNS data set. The data is from the census bureau, and Data-Miners used it as a companion set to Berry and Linoff's book (2004). They used it to build a predictive model; however, we will use it to cluster towns together. Before we attempt this, however, we will need to explore the world of cluster algorithms again as we did in Chapter 3, "Distance: The Basic Measures of Similarity and Association."

6.2 Representing Many Attributes in Multi-dimensions

Back in Chapter 3, we looked at measuring distances by using an inner product between two vectors; see Figure 3.3. Imagine that you have several variables of differing types, such as categorical versus numeric, and even the numeric variables have widely varying meanings such as revenue and age, for example. Although age and revenue are both numeric, the units each represents are very different and the scales could be as well. Age could be in years, decades, or half-year intervals, etc. Revenue could be in units of dollars and the scale perhaps in single dollars or thousands, millions, billions, etc. When you measure the distance of one customer to another in revenue (dollars), and another distance metric like age in years, your measures are in widely differing scales and units, and they cannot be used to compare one distance metric versus another. One method of placing both revenue and age on the same scale is to transform the numbers so they have the same set of units and scale. For example, if dollars and years could be transformed onto a scale that goes from zero to one, then both could be compared with one another. We discussed scaling methods with a few common formulas in Table 3.4. The clustering node in SAS Enterprise Miner will compute scaling if the internal standardization is set to something other than *none*. If you select none, then you can perform your own standardization for the desired set of variables and then when the clustering node is run, your own inputs will be directly sent to the cluster algorithm without any scaling. If you select range standardization, then SAS Enterprise Miner will divide your values by their range to perform the scaling transformation.

One method of scaling not mentioned in Table 3.4 is the SOFTMAX function. It will take a variable such as revenue or age and transform the values to a scale from zero to one (Pyle 1999, pp. 271–274, 355–359). This type of data transformation is needed especially when the data spans many magnitudes. For example, revenues for customers could span anywhere from 0 to 300,000. Let's say we have a range of revenue numbers between 3 and 300,000. If these numbers are expressed in powers of 10, then 3 becomes 3×10^0 and 300,000 becomes 3×10^5. The number 10 when raised to the 0 power becomes 1, so these two numbers expressed as powers of 10 span 5 orders of magnitude. Range scaling is a typical reason to use a function such as softmax. Other reasons include the desire to express data as probabilistic entities (data values from 0 to 1 that also sum to 1).

The SOFTMAX function is shown in Equation 6.1. A SAS macro, which computes the softmax transformation, is shown in Figure 6.1. This SAS macro is available on the data CD in the Chapter 6 folder. In SAS 9.1, there is a CALL SOFTMAX routine as well. Later in Chapter 10, "Product Affinity and Clustering of Product Affinities," we will compare the SAS SOFTMAX function to the one in the following macro when we review one method for dealing with product quantity data.

$$X_n = \frac{1}{(1 + e^{-X_t})}$$

$$\text{and } X_t = \frac{(X_i - \bar{X})}{\lambda(\sigma_i / 2\pi)}$$

(6.1)

where X_t is the transformed value of X_i,

σ_i is the standard deviation of the variable X,

λ is the linear response in standard deviations,

and π is approximately 3.14.

The first part of Equation 6.1 is what is typically referred to as a *logistic function*. The second part scales the linear portion of the logistic function.

Figure 6.1 SAS Macro to Compute the SOFTMAX Function

```
/* Macro Softmax Transform                                      */
/* This macro computes one of three scaling transforms, linear, */
/* unscaled softmax, and scaled or squashed softmax.            */
/* The confidence level used is 90% which is a normal z score of */
/* about 1.283 which is fixed in this application.              */
/* log=L for linear scale, log=S for unscaled softmax, log=SS for */
/* squashed-scaled softmax.                                     */
%macro softmax(dsin=,var=,dsout=,log=L);

 proc sql;
   create table work.stats as
   select min(&var) as minv,
          max(&var) as maxv,
          mean(&var) as meanv,
          std(&var) as stdev
     from &dsin
 ;
quit;
   data _null_ ;
     set work.stats;
       call symput('minv',minv);
       call symput('maxv',maxv);
       call symput('meanv',meanv);
       call symput('stdev',stdev);
    run;
%if %upcase(&log)=L %then
    %do;
        data &dsout;
        set &dsin;
         sm_&var = (&var - &minv)/(&maxv - &minv);
        run;
    %end;

  %else
    %if %upcase(&log)=S %then
        %do;
            data &dsout;
              set &dsin;
              %let var1 = (&var - &meanv)/(1.283 *
```

```
(&stdev/6.2831853));
              sm_&var = 1/(1 + exp(- &var1));
          run;
       %end;
  %else
    %if %upcase(&log)=SS %then
       %do;
          data &dsout;
            set &dsin;
             %let var1 = (&var - &meanv)/(1.283 *
(&stdev/6.2831853));
             %let var2 = 1/(1 + exp(- &var1));
                sm_&var = &meanv +
                ((1.283 * &stdev*log(sqrt(-1+(1/(1-&var2)))) )/
3.14159265 );
            run;
      %end;
%mend softmax;
```

You can use the SOFTMAX function, or something similar, to transform some of your numeric variable ranges from zero to one. This type of transformation will allow a common distance measurement across all scaled variables whether they are numeric or character, ordinal or interval, or even if they have widely varying units of measure. After this type of scaling computation, a cluster algorithm can then find similar groups of observations (records or rows in a data set) on variables that are now more like each other. In addition, when the scaled variables are transformed back into their original units, the profiling of these variables will help in explaining the groups of clusters that will allow you as an analyst to write a description of each cluster's most distinguishing features. When the number of variables is very large, on the order of several thousand variables, you may need to understand which set of variables is most useful in the analysis to follow prior to variable scaling. We will attempt this technique in the next mining example in order to see if the number of variables can be reduced to a more manageable set, but will adequately explain the data. This is what is often referred to as a *parsimonious model*, one that has the fewest number of variables and combinations but explains the data the best.

The dimensionality of a data set is really related to the number of variables it has. One might ask why we can't just feed the data mining algorithm all the variables and just see what comes out? Well, you have probably heard of the old phrase "garbage in, garbage out" and this is what will happen in general if you attempt this kind of analysis. Techniques such as principle components are designed to reduce the number of dimensions (variables) in a data set; however, this has some serious drawbacks. Principle components deals with variables that, in general, have linear relationships with each other. This technique should be applied carefully because it can destroy any non-linear relationships, if they exist (Pyle 1999, pp. 271–274, 355–359). Another problem with many dimensions (e.g., high dimensionality) is no matter how fast or powerful a computer is or what type of data mining software is used, there are always a number of dimensions that will overwhelm any effort at constructing a comprehensive model. Another similarly related problem is the number of possible combinations that arise in all of this data. This problem is caused by the number of unique values increasing as the number of levels of each variable. For example, in a data set there are three variables, each with 3, 4, and 5 unique levels or values respectively. Then the number of possible combinations in these three variables is 3 x 4 x 5 = 60 possible combinations. If the variables in a data set has tens, hundreds, or even thousands of unique levels, the number of possible combinations gets very large very quickly. This is what is sometimes referred to as the *curse of dimensionality*.

There are sometimes other difficulties with many variables, resulting in many dimensions. The problem is typically one of data representation rather than high dimensionality (Pyle 1999, pp. 271–274, 355–359). Representing data in many dimensions can alter the distance metrics used when data is transformed back into its original set of units and values. In other words, clustering data that has been transformed may not represent those same clusters when transformed back. This is one of the main arguments for performing many transformations and has led researchers to develop cluster algorithms that are not sensitive to abnormality and scaling. However, these algorithms are usually very compute intensive, sometimes so much so that they are impractical for most computer systems in a business environment. Yet, as computer hardware continues to progress in speed and agility, algorithms that are more intensive can be written and used with a higher degree of frequency. Figure 6.2 demonstrates this issue of data representation graphically (Duda, Hart, and Stork 2001, p. 573). The figure shows how a distance metric (Euclidean in this example) is represented after dimension reduction from three dimensions down to two. Although it would be impossible to draw graphically, if you were to extend the 3 dimensions in Figure 6.2 to say 200 dimensions, then the problem of data and dimension reduction while keeping variable relationships in tact is a huge undertaking.

Another issue is if one variable is exactly represented by another, it is a simple matter to just use one of the variables and leave the other one out of the model. What happens when the same set of information in two or more variables is only partially duplicated? If you drop one of the variables that is partially duplicated, some information loss will take place. These issues and others are why data mining with many fields and many records or rows of data can be a difficult task.

Figure 6.2 Translating Three into Two Dimensions

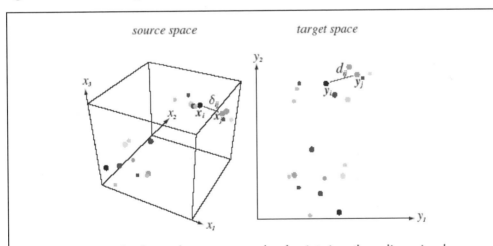

FIGURE 10.26. The figure shows an example of points in a three-dimensional space being mapped to a two-dimensional space. The size and color of each point x_i matches that of its image, y_i. Here we use simple Euclidean distance, that is, $\delta_{ij} = \|x_i - x_j\|$ and $d_{ij} = \|y_i - y_j\|$. In typical applications, the source space usually has high dimensionality, but to allow easy visualization the target space is only two- or three-dimensional. From: Richard O. Duda, Peter E. Hart, and David G. Stork, *Pattern Classification*. Copyright © 2001 by John Wiley & Sons, Inc.

6.3 How Can I Better Understand My Customers of Many Attributes?

In this next example, we will be looking at how to reduce the number of variables to a more manageable level. In typical business situations, the domain experts (experts in the business problem area) will at times dictate some set of specific variables that are desired to keep in the model for the business need rather than for a statistical standpoint. The data set NYTOWNS is comprised of census data from about 1000 New York state towns. The business problem in this example is that a market advertising manager would like to advertise in the state of New York; however, it is too costly and time consuming to come up with 1000 different offers for the consumers of each of the 1000 towns. Your task is to cluster the towns into groups of towns that are similar to each with respect to the variables available in the data set and to profile these clusters so that an advertising campaign can be written for each town cluster. The rules for this example are that no more than twenty different advertising creative media can be developed and no fewer than four should be written for this campaign program. This places the bounds on the segmentation from four to twenty.

So, let's jump right in and start off the exercise with a data assay similar to what was discussed in Chapter 2. This will enable a good understanding of the variables that exist and some of the relationships, missing data, ranges, and the like for this data set. Ensure that the NYTOWNS data set is copied from your CD in the Chapter 6 folder to the SAMPSIO library area. **Step 1:** Open a new project called NY Towns. In the Data Sources node, add a new data source. Follow the instructions as before and you can select the advanced rather than the basic advice in the data advisor selection. Now, create a new diagram and call it NY Towns Clustering. Before we get into profiling the data, let's review some of the groups of variables and their basic attributes on this data set. Appendix 1 shows the complete list of variables sorted by alphanumeric variable name. The label is the description of each demographic variable. The following Process Flow Table shows the steps for this example.

Process Flow Table: NY Towns Clustering

Step	Process Step Description	Brief Rationale
1	Start SAS Enterprise Miner and create a new Data Mining project.	
2	Perform Data Assay on the IncMedFamily and IncPerCapita fields.	Understand potential correlations among variables in NYTOWNS data set.
3	Plot the IncMedFamily and IncPerCapita fields in the Input Data Source node.	Understand potential correlations among variables in NYTOWNS data set.
4	Continue the Data Assay using the StatExplore node.	
5	Observe variable importance with respect to the target variable using the StatExplore node.	Understand which variables are most correlated to the target variable selected.
6	Use the Variable Selection node to minimize the variable list for analysis.	Select only variable with largest importance with respect to Penetration.

(continued)

Process Flow Table (*continued*)

Step	Process Step Description	Brief Rationale
7	Understand the results from the Variable Selection node output.	Discover what the variable selection analysis found.
8	Add a Cluster node and connect the Variable Selection node output.	Feed the results of var-select into the clustering algorithm for analysis.
9	Adjust the clustering node options for another iteration.	Modifying cluster options to obtain a more uniform set of towns in each cluster.
10	View the cluster distance plot on the second pass clustering.	Cluster distance plot helps to see relative size and position of clusters with respect to each other.
11	Perform additional cluster profiling plots.	Understand what makes up each cluster.

As can be seen in Appendix 1, there are many variables to select from, a total of 249. This exercise is to aid in selecting sets of variables for the particular analysis at hand. Many of these variables are related to one another, as they are very similar. For example, the variables IncMedFamily and IncPerCapita are most likely correlated to each other. **Step 2:** To observe the correlation, let's plot these two variables. With your Input Data Source node in your diagram, you can do this easily. Click and highlight the Input Data Source node for the NYTOWNS data set, and in the properties area click on the Variables selection. When the Variables window opens, you can select the **Explore** button, which will allow you to do some simple exploratory analysis. Scroll down the Variables window to the two variables of interest and highlight each variable as shown in Figure 6.3.

Figure 6.3 Highlighting Variables in the Explore Window

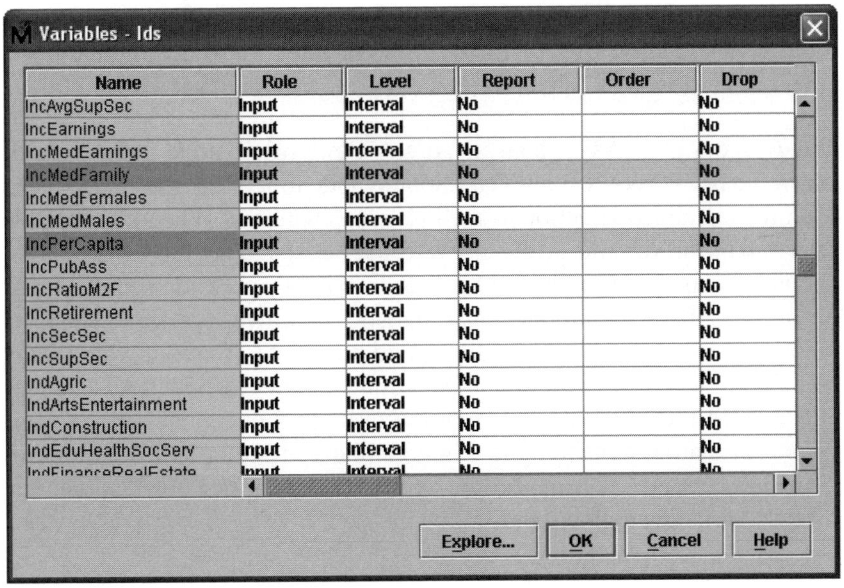

Step 3: Now click the **Explore** button and you should see two histograms of these variables. Click on the Plot selection, select Scatter as your option, and you should see a scatter plot of these two fields as shown in Figure 6.4.

Figure 6.4 Scatter Plot of Variables IncPerCapita and IncMedFamily

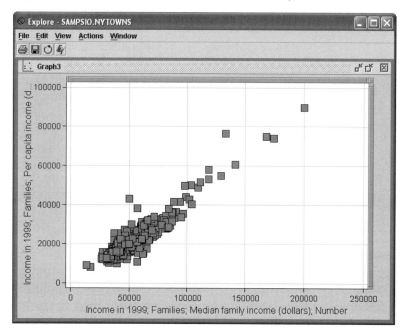

6.4 Data Assay and Profiling

It doesn't take too much statistics knowledge to realize that the plot in Figure 6.3 indicates that these two variables are highly correlated to one another. You should do a few more variables on your own; I would recommend taking one variable from each major group (select Income in 1999; Families; Less than $XX rather than all of the income fields as an example) in your profile analysis. This should give you a feel of each major group or category of demographic attributes. Refer to the data field descriptions in Appendix 1.

Step 4: To continue with our Data Assay analysis, drag a StatExplore node to the diagram workspace and attach the NYTOWNS Input Data Source node to it. With the StatExplore node highlighted, click on the Variables selection in the Properties window. When the Variables window opens, select all of the variables (less any rejected ones) and set the USE field to Yes for all non-rejected variables. Also, you might want to set all of the Correlation settings in the Properties window to Yes. This will compute basic statistics on all variables and correlations. Since we are attempting to cluster these observations and we want to find out which set of variables to use, set a target variable in the Input Data Source node (PENETRATION) to the target level. The PENETRATION variable is the proportion of product penetration in each NY town. This will create correlation plots (among others) in the StatExplore node. After these settings, run the StatExplore node. You should see Person and Spearman correlation plots of the target PENETRATION variable as shown in Figure 6.5.

Figure 6.5 Correlation Plots of Penetration Target Variable

SAS Publishing License Agreement

Important – Please carefully read the terms and conditions of this License Agreement ("Agreement") before installing the enclosed software. By installing the enclosed software ("Product"), you, the purchaser and end user ("Customer") agree to these terms and SAS Institute Inc. ("SAS") will authorize access to and use of the requested Product. If Customer does not agree to all of the terms of this Agreement, Customer must return the Product to SAS. If Customer paid for the Product, Customer may obtain a refund by returning proof of payment along with the Product within thirty (30) days from the date of purchase. If Customer received its copy of the Product at no charge as part of the annual license for SAS software, then Customer will not be entitled to a refund because no payment was made for the Product.

1. **License Grant**

Under the terms and conditions hereinafter set forth, SAS grants to Customer a nonsublicensable and nonexclusive license to access and use the Product as follows:

Customer may copy the Product to a single hard disk of computer hardware that is owned or controlled by Customer for Customer's personal use. Customer may also print pages of the Product for the individual use of Customer. Customer may not load the Product on either an external internet or an internal intranet server. Customer must retain the original media containing the Product during the term of this Agreement. Customer may not duplicate the original media containing the Product except as necessary for backup purposes. Customer may transfer the Product and this Agreement to another party, provided that the other party agrees to accept the Terms and Conditions of this Agreement. If Customer transfers the Product to another party, this license for access to and use of the Product will transfer to the other party and Customer will not retain any rights with respect to access to or use of the Product and must destroy or remove from all hard drives, networks, and any other storage media all copies of the Product, retaining no portions thereof.

If you received the Product free of charge with your SAS software order, your Master License Agreement with SAS shall supersede and replace this Agreement.

ı plot of each variable's
rth.

Move your cursor over the plotted line in Figure 6.6 and a small pop-up window should appear to indicate which variable you are selecting. The upper left-most variable is shown in an image in Figure 6.7 with the mouse pointing to the variable HouseMultiFamily.

Figure 6.7 Inset of Figure 6.6 Showing Variable Importance of HouseMultiFamily

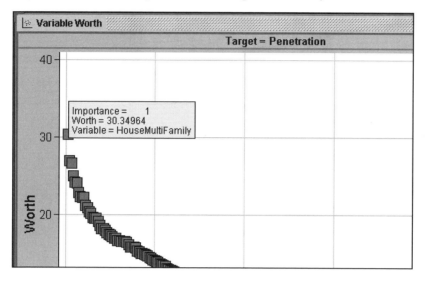

The top several variables are indicating that these contribute the most with respect to the target variable PENETRATION. Although, we are not really looking at trying to predict the PENETRATION variable, we are using the PENETRATION variable as a *proxy* in order to profile a set of variables that could potentially find a group that will be useful in clustering the towns. There are other analyses, basic statistics, and plots to review using this tool; however, let us see if we can limit the set of variables. The SAS/STAT VARCLUS procedure attempts to divide a set of variables into non-overlapping clusters, such that each cluster can be interpreted as essentially one dimension. In essence, this means that the one variable selected in a cluster is representative of a larger set of variables and often with little loss of information (Yeo 2003, pp. 2–49). After running the VARCLUS procedure, you would use just one of the variables within that cluster which is most representative of that cluster of variables. You can run this by dragging a SAS Code node and running the VARCLUS procedure on the NYTOWNS data set; see additional problem sets at the end of this chapter for more details.

Step 6: Another method of limiting the variables is the Variable Selection node. So, let's get started with how this variable grouping or clustering can be accomplished within SAS Enterprise Miner. Drag a Variable Selection node to your workspace and connect the Input Data Source node of the NYTOWNS data set to it. Make sure you have the advanced property sheet set instead of the basic. The Variable Selection node has two basic methods depending on the type of target variable. If the target is a class level, then the Chi-Square Options will apply; for interval targets like what we selected in Product Penetration, the R-Square Options will apply. Since we are trying to reduce the number of variables from almost 250 to something more manageable, let's set the Maximum Variable Number to 30. This will keep only the top 30 variables that have some R-square value against the target variable. Also, set Use AOV16 Variables to Yes as this will bin the input variables into 16 equally spaced groups to detect interactions. Set the remainder of the property sheet values so that they match that of Figure 6.8.

Figure 6.8 Property Sheet Settings for the Variable Selection Node

Property	Value
Node ID	Varsel
Imported Data	...
Variables	...
Max Class Level	100
Max Missing Percentage	50
Target Model	Default
Hide Rejected Variables	Yes
Reject Unused Variables	Yes
⊞ Chi-Square Options	
⊟ R-Square Options	
Maximum Variable Number	30
Minimum R-Square	0.0050
Stop R-Square	5.0E-4
Use AOV16 Variables	Yes
Use Group Variables	No
Use Interactions	Yes
SPDS	Yes
⊟ Status	
Last Error	
Last Status	Complete
Needs Updating	No

Now you can run the Variable Selection node and when you view the results, you should see an R-square bar chart and a table that lists which variables were included and which ones were rejected. Also, you can select the option in the results as:

View→R-Square: Plots→Effects in Model

in the menu and plot the relative importance of variables selected to the target. These are shown in Figure 6.9.

Figure 6.9 Variable Selection Node Results Windows

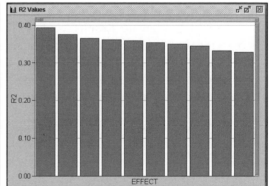

Name	Role	Level	Type	Label
AncItalian	Input	Interval	N	ANCESTRY (single or mu.
AOV16_HouseMultiFamily	Input	Ordinal	N	AOV16: HouseMultiFamily
EduHSDip	Input	Interval	N	Educational attainment; P.
IndAgric	Input	Interval	N	Employed civilian populati.
JobAgriculture	Input	Interval	N	Employed civilian populati.
JobOfficeSales	Input	Interval	N	Employed civilian populati.
MarFemaleDivorcees	Input	Interval	N	% of marriage aged femal.
MortageLT500	Input	Interval	N	Percent of mortgages wit...
MortgageLT700	Input	Interval	N	Percent of mortgages wit...
ValueLT50K	Input	Interval	N	% value less than $50,000
AncArab	Rejected	Interval	N	ANCESTRY (single or mu.
AncCzech	Rejected	Interval	N	ANCESTRY (single or mu.

(continued)

Figure 6.9 (*continued*)

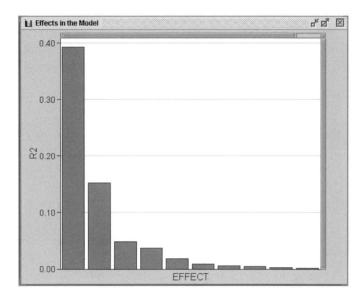

Step 7: Place your mouse over each bar to see what variable is represented. The largest bar is the AOV16:HouseMultiFamily. This means that after dividing the HouseMultiFamily variable into 16 equally spaced sections, this *new* variable is the most important with respect to the other variables in the model. If we did not need any other variables from a business (rather that statistics) perspective, then we could just take these variables and reject the others. So let us attempt this by dragging a Metadata source and connecting the Variable Selection to it. Then, change the target variable (Product Penetration) to an Input rather than a Target.

Step 8: Now connect a Cluster node from the output of the Metadata node. Only the variables that were identified in the Variable Selection node will be passed to the Cluster node. Set the maximum number of clusters to 20 to satisfy the marketing manager's requirement. Let's leave everything else at the default value and see what turns up. Run the Clustering node and when complete, view the Results window. You should see 14 distinct clusters in the Results Pie Chart window. The pie chart called Segment Size in the Results window is shown in Figure 6.10. We should now profile these clusters and determine if any cluster modifications should be made. In addition, you should open the Variable Importance table to see which variables are responsible for the main definition of these clusters. To do that, use the menu selection:

   View→Cluster Profile→Variable Importance

Figure 6.10 Initial Clustering Segment Size Plot

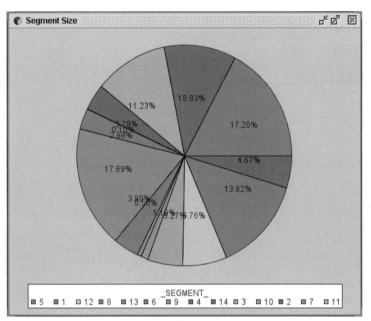

Step 9: Notice in the Mean Statistics table that one of the clusters has only one town in it, as seen in Figure 6.11. Viewing the size of each cluster and its variability may not be useful as cluster segmentation in that the variation in the number of towns in each cluster is quite high. The marketing manager would probably like something more uniform if possible. So, without continuing to profile this clustering iteration, let's make a few modifications to fix this problem first.

Figure 6.11 Initial Cluster Mean Statistics Table

SEGMENT	Frequency of Cluster	F
1	173	
2	109	
3	113	
4	38	
5	1	
6	30	
7	178	
8	40	
9	5	
10	12	
11	53	
12	68	
13	139	
14	47	

The default settings always set the Standardization to None. This means that when different variables are compared, they are compared on their original scales and as we discussed earlier, this is a problem for distance measurements. So, change the Internal Standardization to Range and run the Clustering node again. You should now notice that the number of clusters is far fewer, five in this pass, and the sizes are a bit more uniform with the slight exception of cluster 1 with 71 towns. Figure 6.12 shows the revised clustering towns per cluster and the variable importance table.

Figure 6.12 Second Pass Cluster Mean Statistics and Variable Importance

Mean Statistics

ne...	SEGMENT_	Frequency of Cluster
.	1	71
.	2	181
.	3	155
.	4	324
.	5	275

Variable Importance

NAME	LABEL	RELATIVE IMPORTANCE
Penetration	Product penetration (percent of households)	1
EduHSDip	Educational attainment; Population 25 years and over; ...	0.987488
AncItalian	ANCESTRY (single or multiple); Total ancestries report...	0.936264
MortgageLT700	Percent of mortgages with payment less than $700	0.859119
ValueLT50K	% value less than $50,000	0.821834
IndAgric	Employed civilian population 16 years and over; Industr...	0.730758
MortageLT500	Percent of mortgages with payment less than $500	0.716715
AOV16_HouseMult...	AOV16: HouseMultiFamily	0.703609
MarFemaleDivorce...	% of marriage aged females who are divorced	0.607306
JobAgriculture	Employed civilian population 16 years and over; Occup...	0.410377
JobOfficeSales	Employed civilian population 16 years and over; Occup...	0.329223

6.5 Understanding What the Cluster Segmentation Found

The variable importance shows that the most important distinction between clusters of towns is the product penetration. This is to be expected since we selected variables that relate to product penetration in the previous analysis. In fact, if the PENETRATION variable was not at or near the top, perhaps we did something wrong. Next most important is the educational attainment of people 25 years and older. You can see this readily by scrolling over the Mean Statistics table and looking at the product penetration column. This will show the average product penetration for each cluster. Note that the product penetration varies from 1.8 to 19 in the five cluster segments and at the same time, the educational attainment varies from 42 to 29. Segment 1, therefore, has the highest product penetration and educational attainment with people 25 years and older.

Step 10: One of the next items you might want to view is the Cluster Distance Plot. Part of profiling the cluster segments is the ability to visualize the structure of our multidimensional data. The problem is, however, that we don't know how to plot a graph in 12 or 15 dimensions! So, one way to attack this problem is to try to represent the data in the clusters as two dimensions so that these distances correspond to the similarities of data points in the original set of dimensions. If a reasonably accurate representation can be found in two or perhaps three dimensions, then this can be a valuable way to gain insight into the structure of the data visually (Duda 2001, p. 573). The Cluster node in SAS Enterprise Miner performs multidimensional scaling (PROC MDS) on the distance metrics and plots the dimensions that have the largest amount of variability; that is the two dimensions that have the largest eigenvalues. Figure 6.13 shows the Cluster Distance plot. In the Results window there is an Output window where you can find the set of eigenvalues that were computed for the covariance matrix in the clustering algorithm computations. Table 6.1

shows the eigenvalue estimates. The distance plot of dimensions 1 and 2 uses the largest two eigenvalues. The proportion indicates that about 89% of the variability is explained by the first two eigenvalues! One or more variables make up a dimension; therefore, it is necessary to profile what each of those dimensions is attempting to explain. In Figure 6.13, cluster 5 must be rather different from cluster 2 in the dimension 1 direction, and cluster 1 must be rather different from cluster 3 or 2 in the dimension 2 direction.

Table 6.1 Eigenvalue of the Covariance Matrix

	Eigenvalue	Difference	Proportion
1	387.418941	351.136253	0.8299
2	36.282687	21.856238	0.0777
3	14.426449	2.180802	0.0309
4	12.245647	5.672905	0.0262
5	6.572742	0.466032	0.0141
6	6.106709	3.701251	0.0131
7	2.405458	1.281513	0.0052
8	1.123945	0.925953	0.0024
9	0.197992	0.167073	0.0004
10	0.030919	0.029777	0.0001
11	0.001141		

Figure 6.13 NY Towns Cluster Distances Plot

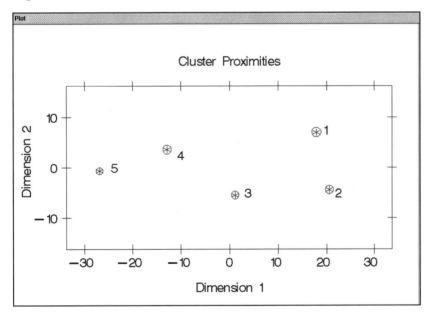

Figure 6.13 indicates that the clusters are reasonably well separated and distinct; the tight circles indicate that there is rather good uniqueness of each cluster. If the cluster circles were to overlap a good deal, then the distances from the cluster centroid would be somewhat large and this might be indicative of observations that could possibly fall into more than just one cluster. In the case here, we are using *disjoint clustering* so each and every observation will fall into just one cluster. In *fuzzy clustering*, the probability of cluster membership is typically used as a metric, and the possibility of multiple cluster memberships is allowed. The topic of fuzzy clustering will be discussed a bit more in Chapter 11, "Computing Segments Using SOM/Kohonen for Clustering."

Step 11: Another area for profiling while in the Cluster Node Results window is the Segment Plot window. The segment plot shows each variable and the percentage that is found within each cluster segment. Figure 6.14 shows that in segment variable 5, the larger bar represents 94% where product penetration is between the formatted range of 0 through 6.0625%, and the remaining 5.8% is between 6 and 12%.

Figure 6.14 Segment Plot Window of Product Penetration

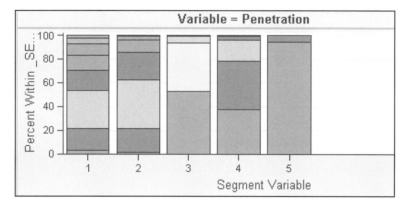

Therefore, segment 5 does not have very large product penetration. This can also be seen in the averages of the Mean Statistics Table window; it shows that cluster 5 has an average product penetration of 1.8%. Contrast this with cluster 1, which has an average product penetration of 19.97% and from the segment plot, about 50% of cluster segment 1, has product penetration above 25 to 30%. Clusters 5 and 1 are on opposite ends of the product penetration spectrum, and product penetration was the largest contributor in the variable importance; referring to Figure 6.13, we can infer that dimension 1 is most likely made up of product penetration. In this manner, we can continue to profile each cluster segment and determine the spatial relationship from the distance plot. It is sometimes helpful to actually edit the distance plot, place on the axis a set of variable representations for each dimension, and also label the clusters so one can easily visualize the main differences between the clusters. You could also generate a tabular report that gives the main profile elements for each cluster and perhaps a name that reflects that cluster's attributes. Cluster 1, for example, might be aptly named High Penetration Towns. Cluster 1 also ties with cluster 2 for average educational attainment so you could also label cluster 1 as High Edu-Penetration Towns.

6.6 Planning for Customer Attentiveness with Each Segment

At this point in the analysis, when the cluster segments are fully profiled, a question of "what next" might be typical thought. Now that there are five distinct segments in which all of the towns will be classified, it will be up to that marketing manager to decide what kind of creative advertising should be developed for each cluster segment. Once the advertising is created, then all the towns in each segment would receive the creative media. Obviously, additional cluster segments could be found, and one method of doing this is to take the towns just in one of the cluster segments, and perform the same type of variable reduction and clustering within that one segment to see if other sub-clusters emerge that could potentially help in the marketing manager's campaign. I'm not going to perform this exercise here, but you should attempt it at some point. This specific exercise is listed in Section 6.8 at the end of this chapter.

The next question might be now that we have cluster segments and their profiles, what do we do with them? This kind of question is at a pivotal point that can turn the analysis from just an exercise into something that is actionable in business. Segmentation, clustering, or any other data mining activity has absolutely no value if nothing actionable comes from the analysis. An action might be that we need more information, but some action must result from the analysis. If the intention of the analysis is *to better understand* my base of customers, then once that understanding is available, it should generate some action that results in a change of business practice, or confirms or denies some hypothesis made about the customer base. Whatever the situation might be, it should always result in some action or the analysis is of very little value indeed. In the previous example, the marketing manager with five cluster segments to start from can plan for the types of advertising in each of the segments.

If this data set of NY towns was really a much larger data set, say instead of 1,000 towns it contained 3 million (pretty unlikely for the state of NY), then could the same analysis take place? What if the size of the data set were 120 million customer or prospect data records? Would you really want to place 120 million data records into a cluster algorithm? This is the topic of our next section.

6.7 Creating Cluster Segments on Very Large Data Sets

To answer questions relating to clustering of vary large sets of data, we would need to qualify them with some additional information. Let's say that you have a moderate size server with adequate memory and generous data storage for the 120 million-record data set. Now, would it be advisable to perform the clustering technique we just did for all 120 million records? The answer might be probably not. Why? Because it may take the server many hours to run that kind of algorithm and for practical development of cluster segmentation, several iterations are typically required and thus the turnaround time for each pass through the algorithm needs to be much faster. One possible solution would be to run the clustering algorithm on a *statistically valid* sample of the original data set that is much smaller in the number of records. Then if you can create a set of scoring code that will perform the computations of the clustering solution on the remaining data set, a complete solution arises. This method takes much less time, and the iterative process of cluster analysis development is not the bottleneck for accomplishing the final solution on the larger data set. This technique just mentioned is available through SAS Enterprise Miner. The Scoring node will take inputs from the Cluster node as well as the modeling nodes. To see this, open the last diagram in your project called NY Towns. Add into your diagram a Score node and connect the Cluster node to it. In the Score node property sheet, click the **Imported Data** button and you should see in the new window the Cluster node data set that is a predecessor to the Score node. Set the Type of Scored Data to Data instead of View for this case. The Imported Data window is shown in Figure 6.15.

Figure 6.15 Score Node Imported Data from the Clustering Node

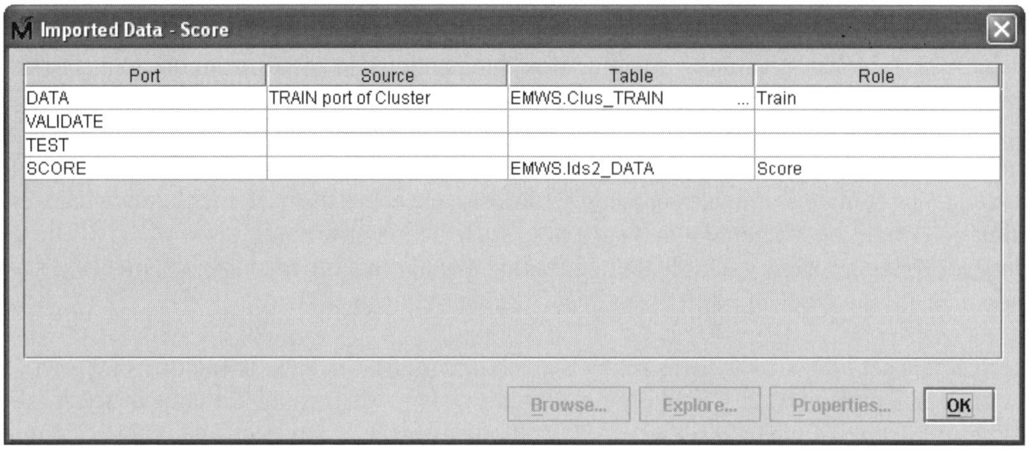

When you run the Score node, the node will generate C, Java, and SAS code. If you want to score your data in another SAS data mart, then when the Score node is complete, open the Results window of the Score node and select **View→SASCode→Path Publish Score Code** to view the actual SAS code. See Figure 6.16.

Figure 6.16 Results Window of the Score Node: Viewing SAS Code

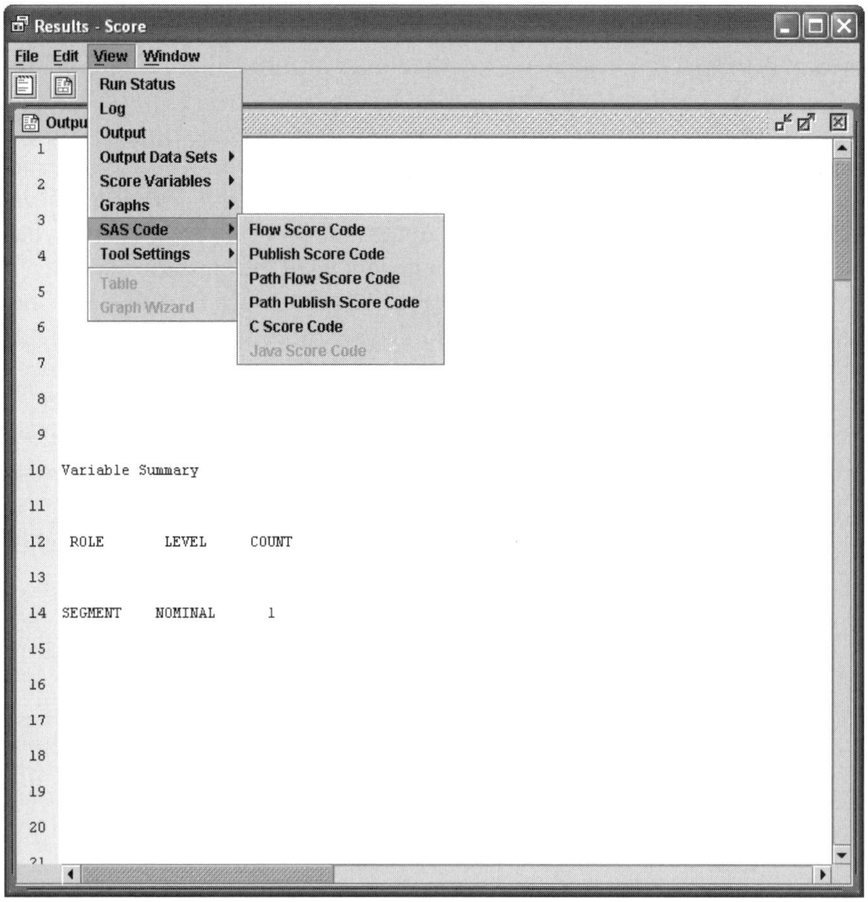

This code can easily be converted to a SAS macro by copying all the SAS code in the view and then editing it to include a macro header and ending statement and *data* and *set* SAS statements as shown in Figure 6.17. This macro can then be imported to another system that contains an entire data warehouse or data mart and as long as the same set of variables are in that data environment. This code will score that data set with the clusters from the clustering model. It should be noted here that this is not the same as actually performing a clustering on the data; it is scoring only according to the model that was built. So in order for the model to apply, the data set that the model was built from must be statistically representative of the data that is being scored; otherwise, the model scores will not be applicable.

Figure 6.17 Viewing SAS Code and Creating a SAS Macro

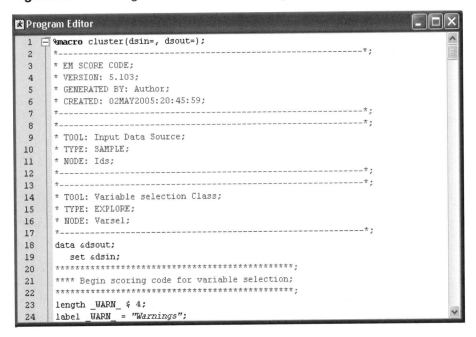

```
Program Editor
 1   %macro cluster(dsin=, dsout=);
 2   *-----------------------------------------------------*;
 3   * EM SCORE CODE;
 4   * VERSION: 5.103;
 5   * GENERATED BY: Author;
 6   * CREATED: 02MAY2005:20:45:59;
 7   *-----------------------------------------------------*;
 8   *-----------------------------------------------------*;
 9   * TOOL: Input Data Source;
10   * TYPE: SAMPLE;
11   * NODE: Ids;
12   *-----------------------------------------------------*;
13   *-----------------------------------------------------*;
14   * TOOL: Variable selection Class;
15   * TYPE: EXPLORE;
16   * NODE: Varsel;
17   *-----------------------------------------------------*;
18   data &dsout;
19       set &dsin;
20   ***************************************************;
21   **** Begin scoring code for variable selection;
22   ***************************************************;
23   length _WARN_ $ 4;
24   label _WARN_ = "Warnings";
```

6.8 Additional Exercise

Create a Data Mining process flow in addition to the five cluster segments found earlier in this chapter, and process each cluster segment to see if any other sub-clusters emerge. Compare and contrast each sub-cluster profile to its top-level cluster profile (e.g., if 3 sub-clusters were found in cluster 1, then compare those sub-cluster profiles to cluster 1's profile). Do any additional information and use in the marketing program arise, or are the sub-clusters just further definitions or explanations of the top-level cluster?

6.9 References

Berry, Michael J. A., and Gordon S. Linoff. 2004. *Data Mining Techniques.* 2d ed. New York: John Wiley & Sons, Inc.

Berry, Michael J. A., and Gordon S. Linoff. 2004. *Data Mining Techniques.* 2d ed. New York: John Wiley & Sons, Inc. Companion pages that contain data sets and chapter presentations for this book are available at http://www.data-miners.com/companion/dmt.html.

Duda, Richard O., Peter E. Hart, and David G. Stork. 2001. *Pattern Classification.* New York: John Wiley & Sons.

Pyle, Dorian. 1999. *Data Preparation for Data Mining.* San Francisco: Morgan Kaufmann Publishers, Inc.

SAS Enterprise Miner, Release 5.2. Cary, NC: SAS Institute Inc.

Yeo, David. 2003. *Applied Clustering Techniques Course Notes.* Cary, NC: SAS Institute Inc.

When and How to Update Cluster Segments

7.1 What Is the Shelf Life of a Model, and How Can It Affect Your Results?

When you have come up with a satisfactory cluster segmentation model and profiled each of the cluster segments, one practical question that naturally arises is how long this model will last until you need to completely re-cluster any new observations added to the original data set. Alternatively, to rephrase the question a little, can you use the same cluster scoring model on a data set even with new records, and how can you tell that the clustering model needs to be refitted? The answer to questions like these is the topic of this chapter, and I should say at the outset that there has not been an abundant supply of literature on this topic, to say the least. Much work has gone into various algorithms for clustering, especially when the data is ill-behaved or for specific applications such as image clustering or clustering of textual document data, but there has not been too much research on the topic of practical applications of when cluster algorithms need to be refitted when there are new data records. Shelf life of a cluster model is basically when your model no longer performs satisfactorily on new sets of observations and requires that the model be refitted. A key question to think about is what is considered satisfactory? We will come back to this question in a little while.

To get a better idea of the process involved in refitting a clustering algorithm, we should briefly revisit the main measurements and criteria of clustering. In Chapter 3, "Distance: The Basic Measures of Similarity and Association," the technique that brings about how clusters are evaluated is based on measuring similarity, specifically a distance or proximity metric between any two or more records in the data set under study. So one of the ways we can evaluate how a clustering algorithm has changed is to measure how distances or proximity metrics have changed on additional observations. If you use the scoring code that SAS Enterprise Miner creates from a cluster model like the last example in Chapter 6, "Clustering of Many Attributes," then this code does in fact generate cluster segments and a distance metric is computed. The score code replicates the completed cluster model that was developed; however, it does not perform any clustering. The score code just implements the existing model. Furthermore, cluster algorithms function by optimizing the data in the training set but do not use a validation data set for additional training like in Neural Network and other models.

When you score additional observations that have not been used in training, then the distance metrics might in fact vary a bit from the originally trained model. Statistically speaking, one could test the distances from the *trained* model against the new records in the test data set of the *scored* model and see if they are significantly different from each other. What is being tested in this scenario is a distance metric in a training set used to perform clustering against that same model (scoring code) scored onto new observations. If these distances are rather close, then the cluster model is scoring the new observations as intended. When distance metrics on new observations do not reflect the same metrics of the older model, then perhaps it is time to refit the clustering model with the new observations and the model development process repeats itself. I will address the issue of how to detect when your clustering model should be updated with refitting the clustering algorithm in Section 7.2.

If an analyst performs a clustering segmentation and uses the scoring code to score new records or observations but never updates the cluster model, eventually the distances of the original clusters may vary so much that the similarity within a cluster may be comparable to the similarity between clusters. This tendency may actually merge the original clusters together so that they might not even become distinguishable.

We might want to review briefly the key question that was raised regarding what is considered satisfactory. When comparing distance metrics on new observations to the ones from the original model, statistically, you might detect that there are really different metrics between the original model data set and the new observations. However, practically speaking, these differences may not be enough to cause alarm and may be perfectly acceptable in the business situation in which the model is being used. Thus, statistics might say there is a real detectable difference; however, the business might say that it can *live* with the difference. These considerations should be taken into account when performing the analysis.

7.2 How to Detect When Your Clustering Model Should Be Updated

So, how might we detect a difference in the cluster model's version of the distances and the newly scored records? As mentioned earlier, comparing the distance metrics on newly scored records to those from the original data set used to develop the cluster model is perhaps one method for detecting when the cluster model should be refitted. Let's give this a go and see if it is possible to detect differences.

Process Flow Table: Distance Metrics

Step	Process Step Description	Brief Rationale
1	Start SAS Enterprise Miner and create a new Distance Metrics project and process flow diagram called Distance Detection.	
2	Add three new data sets to the Data Sources folder of the project.	
3	Add a Transform Variables node. Transform four variables.	Maximizes the normality of four variables.
4	Filter the node to remove outlier observations.	Removes extreme or rare observations from the four variables after transformation.
5	Connect a Cluster node after the filter node.	
6	Add a Score node and the CUST_NEW data set from the Data Sources.	Creates the scoring code from the cluster model and scores new customer records.
7	Use a SAS Code node to summarize simple statistics of newly scored data.	Compares simple statistics of cluster model distances to the newly scored data.
8	Add more code to the SAS Code node to generate histogram plots.	Plots the distances by segment.
9	Add a new data source to the diagram and score the CUST_NEWSCORE data.	Scores another set of data and plotting.
10	Change the second SAS Code node statements for histograms on all 3000 data records.	Allows histograms on all scored records versus just a random sample.

Step 1: Open a new SAS Enterprise Miner project and call it Distance Metrics. **Step 2:** Add three new data sources from the SAMPSIO library: CUST_SUBSET, CUST_NEW, and CUST_NEWSCORE. Create a new diagram called Distance Detection. **Step 3:** Drag the CUST_SUBSET from the Data Sources folder onto your process flow diagram add a Transform Variables node and select the variables TOT_REVENUE, REV_THISYR, EST_SPEND, and LOC_EMPLOYEE; transform theses variables only with a Maximum Normality option.
Step 4: Connect a Filter node and set the default value of Rare values for the filtering method.
Step 5: Now connect a Cluster node and set only the variables to use in the cluster analysis that are shown in Figure 7.1. Ensure that the variable CUST_ID is set to an ID role and that the selection of Yes is included in the listing of variable. The settings in the Cluster node using the advanced property sheet are shown in Figure 7.2.

Figure 7.1 Variables to Use in the Cluster Node

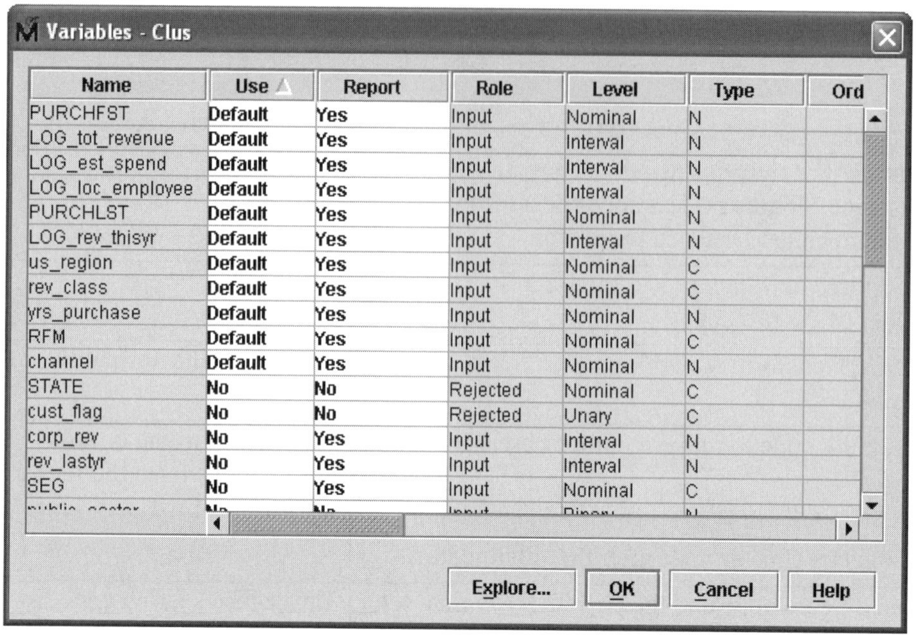

M Variables - Clus

Name	Use △	Report	Role	Level	Type	Ord
PURCHFST	Default	Yes	Input	Nominal	N	
LOG_tot_revenue	Default	Yes	Input	Interval	N	
LOG_est_spend	Default	Yes	Input	Interval	N	
LOG_loc_employee	Default	Yes	Input	Interval	N	
PURCHLST	Default	Yes	Input	Nominal	N	
LOG_rev_thisyr	Default	Yes	Input	Interval	N	
us_region	Default	Yes	Input	Nominal	C	
rev_class	Default	Yes	Input	Nominal	C	
yrs_purchase	Default	Yes	Input	Nominal	N	
RFM	Default	Yes	Input	Nominal	C	
channel	Default	Yes	Input	Nominal	N	
STATE	No	No	Rejected	Nominal	C	
cust_flag	No	No	Rejected	Unary	C	
corp_rev	No	Yes	Input	Interval	N	
rev_lastyr	No	Yes	Input	Interval	N	
SEG	No	Yes	Input	Nominal	C	
public_sector	No	No	Input	Binary	N	

Explore... OK Cancel Help

Figure 7.2 Cluster Node Property Sheet Settings for Analysis

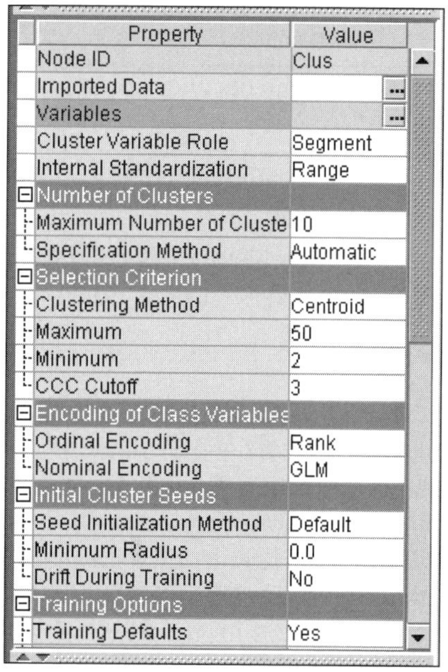

Property	Value
Node ID	Clus
Imported Data	...
Variables	...
Cluster Variable Role	Segment
Internal Standardization	Range
⊟ Number of Clusters	
Maximum Number of Cluste	10
Specification Method	Automatic
⊟ Selection Criterion	
Clustering Method	Centroid
Maximum	50
Minimum	2
CCC Cutoff	3
⊟ Encoding of Class Variables	
Ordinal Encoding	Rank
Nominal Encoding	GLM
⊟ Initial Cluster Seeds	
Seed Initialization Method	Default
Minimum Radius	0.0
Drift During Training	No
⊟ Training Options	
Training Defaults	Yes

Step 6: Connect a Score node from the output of the Cluster node, and drag the data source called CUST_NEW and connect both the Cluster node and the input data of the CUST_NEW to the Score node. Set the role of the CUST_NEW data source to Scoring instead of the default value of Raw. Use this flow when new customer records in the CUST_NEW data set are to scored with the cluster model developed using the CUST_SUBSET data. The process flow diagram should now look like the one in Figure 7.3.

Figure 7.3 Completed Process Flow Diagram of the Cluster Analysis

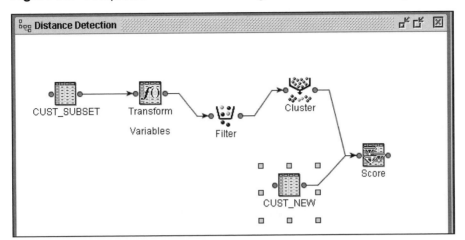

The settings that were chosen should produce a cluster segmentation analysis of six distinct segments. When you've completed the flow diagram and the individual settings, run the diagram either from the Cluster node or the Score node. The Score node will collect all the pertinent transformations, filtering, cluster settings, and the like and combine them into a single code set to be used in C-scoring, Java, or SAS. The final data set from the Cluster node produces two additional variables: _SEGMENT_ and DISTANCE. The _SEGMENT_ variable is the cluster IDs and can be formatted to be any such text suitable for labeling the cluster IDs something other than a numeric label. If you open the Results window from the completed Scoring node, you can view either the training data set or the scoring data set. Figure 7.4 shows the scoring data set opened with the default value of 2,000 randomly fetched rows of data in a partial view of the Results window from the Score node. Note the values of _SEGMENT_ and DISTANCE are the values to be compared to the training data set that was used to build the cluster model.

Figure 7.4 Scored Data Set in the Results Window

ansforme...	Transforme...	Transforme...	Transforme...	SEGMENT_	DISTANCE	Imputed: L...	Imputed: L...
0.006201	0.002109	0.075712	0.126705	6	2.07471	0.006201	0.002109
0.00006	0.000067	0.07753	0.124851	4	2.228162	0.00006	0.000067
0.009138	0.005651	0.075712	0.126943	3	2.40742	0.009138	0.005651
0.000585	0.00111	0.075712	0.123703	3	2.089725	0.000585	0.00111
0.070902	0.00222	0.080348	0.137765	4	1.951386	0.070902	0.00222
0.008529	0.000111	0.075993	0.124216	5	1.938722	0.008529	0.000111
0.000305	0.000711	0.075712	0.12361	3	1.804394	0.000305	0.000711

So, now that the test data set has been scored with the cluster segments and distances, and the training data set also contains segments and distances, how might we compare these two sets? One way might be to summarize the distance metrics with various descriptive statistics (such as mean, standard deviation, range, etc.) for each cluster segment and to compare the two data sets. **Step 7:** To do this, attach a SAS Code node to the Score node and open the Code window by clicking the **SAS Code** tab in the Property Sheet window. Enter the SAS code shown in Figure 7.5. Note the names of the data sets are automated SAS macro variables that point to the specific data sets. SCORE_TRAIN refers to the training data set used in the cluster model; SCORE_SCORE is a macro variable that points to the scoring data set we used to score the cluster segments, and so on.

Figure 7.5 SAS Code Node Window with Summary Statements

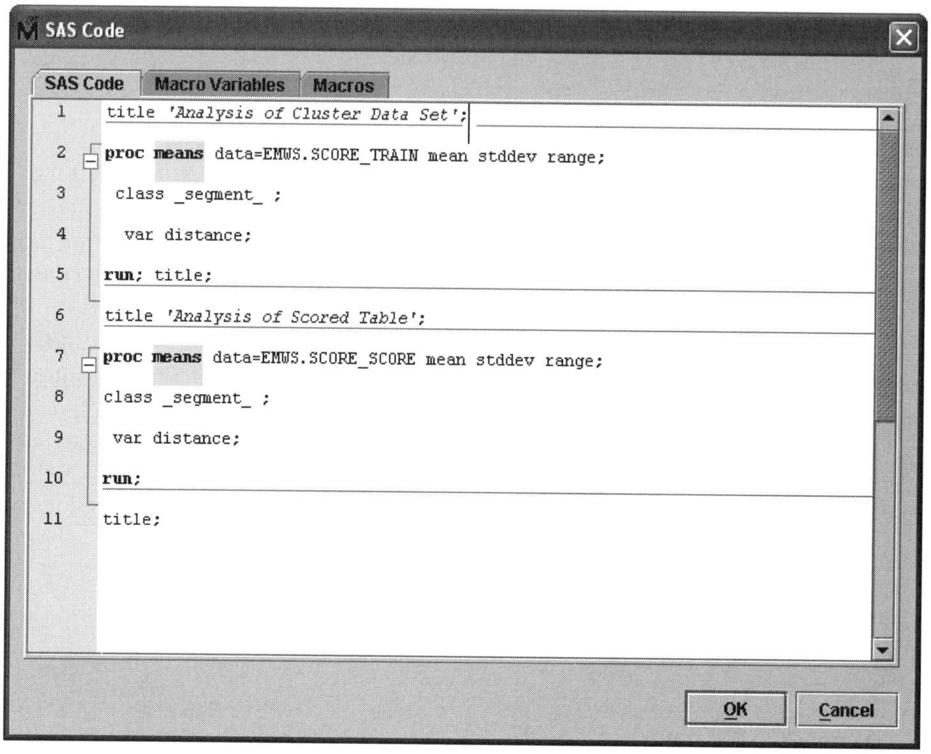

```
1    title 'Analysis of Cluster Data Set';
2    proc means data=EMWS.SCORE_TRAIN mean stddev range;
3      class _segment_ ;
4       var distance;
5    run; title;
6    title 'Analysis of Scored Table';
7    proc means data=EMWS.SCORE_SCORE mean stddev range;
8    class _segment_ ;
9     var distance;
10   run;
11   title;
```

When you run the SAS Code node, the Results window will contain the standard SAS output with the MEANS procedure output from each of the two data sets. The first data set is the one that we used to build the cluster model, and the second is the one in which the scoring code from the model was used to score the data records. The output from the Results window is shown in Figures 7.6 and 7.7. Figure 7.6 shows the results from the scored data using the Scoring node. From these two results tables we can probably conclude that the scored results are nearly identical to the results obtained from the original training set in which the cluster model was developed.

Figure 7.6 Training Data Distances by Cluster

```
Output
22  Analysis of Cluster Data Set
23
24  The MEANS Procedure
25
26                      Analysis Variable : Distance
27
28                  N Obs          Mean        Std Dev          Range
29  ------------------------------------------------------------------
30        1        14638     1.7338475      0.3328823      1.3723940
31
32        2         5140     1.9300521      0.2365963      1.1323579
33
34        3        16204     2.0760855      0.1948554      1.0410849
35
36        4        10169     2.0868862      0.1608226      1.0455987
37
38        5         7308     1.9482254      0.1138427      0.9270864
39
40        6        11944     1.7991883      0.3079381      1.1647135
41  ------------------------------------------------------------------
```

Figure 7.7 Scored Data Distances by Cluster

```
Output
42  Analysis of Scored Table
43
44  The MEANS Procedure
45
46                      Analysis Variable : Distance
47
48  _SEGMENT_    N Obs          Mean        Std Dev          Range
49  ------------------------------------------------------------------
50        1         6247     1.7334050      0.3349469      1.4156635
51
52        2         2278     1.9339222      0.2376068      1.1316854
53
54        3         6852     2.0721076      0.1925602      0.9569053
55
56        4         4326     2.0897273      0.1656631      0.9828529
57
58        5         3123     1.9475385      0.1133430      0.9116918
59
60        6         5013     1.7958778      0.3058392      1.0658570
61  ------------------------------------------------------------------
```

We can also add other means for analysis with graphical aids by including several more statements in our SAS Code node. SAS Enterprise Miner gives you access to write to the model results package. These statements will register the temporary user data sets and access the EM_REPORT macro, which allows you to view data, or to make x-y scatter plots, line plots, or histograms. For more details, see the section on the SAS Code Node Utility Macros in the online documentation. **Step 8:** Open the SAS Code node and add the new SAS statements shown in Figure 7.8 after the last line of code entered as shown in Figure 7.5. When you rerun the entire SAS Code node with the additional statements, you will now see the new drop-down menu Plots in the SAS Code Results window as shown in Figure 7.9. The Plots menu will plot the histograms of the scored data set and the training data set with 10 and 20 percent random sampling using the RANUNI function in the WHERE statement; each plot will contain a histogram of the distances in each cluster segment. These plot comparisons are shown in Figure 7.10.

Figure 7.8 Additional SAS Code for Histograms by Segment

SAS Code

SAS Code | Macro Variables | Macros

```
%em_register(key=tscore,type=data);
%em_register(key=ttrain,type=data);
data &em_user_tscore;
  set emws.score_score(where=(ranuni(0)<=0.2));
run;
data &em_user_ttrain;
  set emws.score_train(where=(ranuni(0)<=0.1));
run;
proc sort data=&em_user_ttrain;
by _segment_;
proc sort data=&em_user_tscore;
by _segment_;
run;
%em_report(key=tscore,viewtype=histogram,x=distance,block=Plots,
by=_segment_,description=Score Data Distance Histogram);
%em_report(key=ttrain,viewtype=histogram,x=distance,block=Plots,
by=_segment_,description=Train Data Distance Histogram);
```

OK Cancel

Figure 7.9 New Menu Selections in the SAS Code Results Window

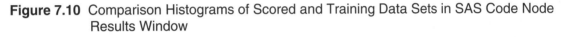

By viewing the histograms by cluster segment, you can easily conclude that the scoring code produced very similar results on the scoring data set as on the training data set. If the scoring data set were new records that were added to your data mart at a later date, for example, then this process flow could aid in the detection of when the distances by cluster segment has changed enough to warrant a refitting of the cluster algorithm.

Figure 7.10 Comparison Histograms of Scored and Training Data Sets in SAS Code Node Results Window

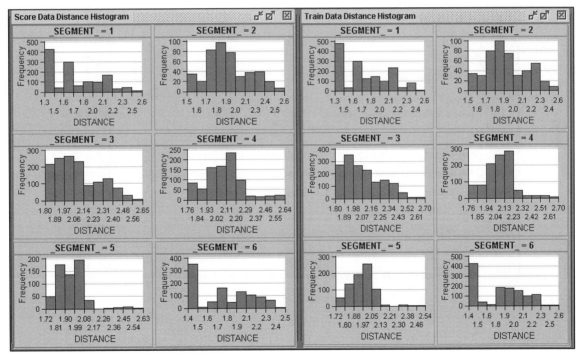

What we can learn from the scored distances versus the training distances is that the scored distributions look *similar* enough to the original distributions, and this can be considered a relatively quick test to see that the segment model is scoring in a similar fashion to what was originally developed. The amount of dissimilarity may be somewhat subjective in deciding that a new model is needed; however, at least you have a metric in which to compare and contrast the amount of similarity that can be agreed upon as a business-level decision.

7.3 Testing New Observations and Score Results

Let's look at what happens to new data records that are to be scored with the clustering model using the SAS Scoring code if the cluster model is not quite the same as the scoring results. **Step 9:** To see this, add in the third data source called CUST_NEWSCORE to your diagram, and set the role of the data set to Scoring. This data set has a little over 3,000 new customer records for scoring. Now click the Score node and right-click the selection Copy. This copies all of the scoring code (the entire node and its settings) in the original and now you can paste it into the diagram. It should be called Score 2 as the second Score node. You can give it a different name if desired. Do the same for the SAS Code node except alter the SAS Code node as shown in Figure 7.11. **Step 10:** Change EMWS.SCORE_SCORE to EMWS.SCORE2_SCORE and remove the RANUNI function for sampling, as we want all 3,000 data records. Now, run the SAS Code 2 node and view the results. You should see in the plot by cluster segment that segments 2 and 5 are somewhat different in shape from segments 2 and 5 in the training set. This plot is shown in Figure 7.12.

Figure 7.11 Revised SAS Code Node 2 to Reflect the New Data Set

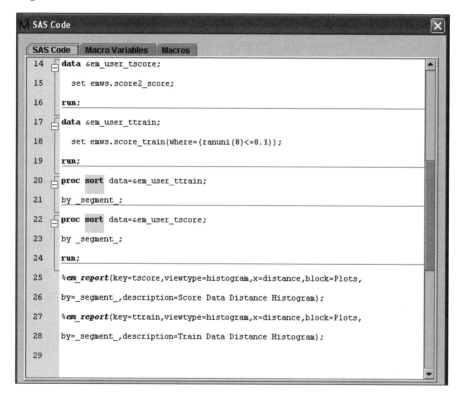

Figure 7.12 Results of SAS Code 2 Node Histogram Plot of Score Data Set
CUST_NEWSCORE

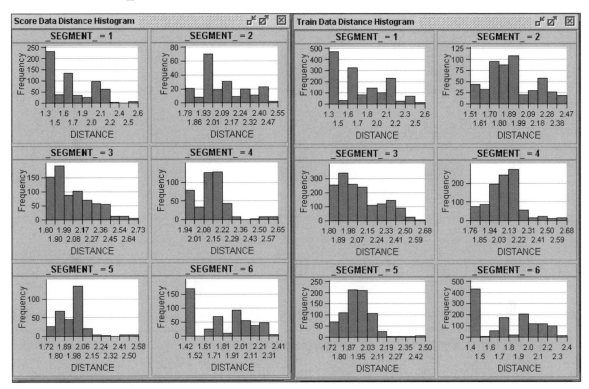

Segments 3 and 4 in the left half of Figure 7.12 are different in shape than segments 3 and 4 in the right half of the figure.

With these small changes, it might be a good idea to re-cluster the new data set with all of the observations from the original data set and the new observations together to come up with a new cluster model. The main question you might want to ask when performing this type of analysis is whether these results change the segmentation if you used the older scoring code or whether you should re-build a new model. If the answer to that question indicates that the results are not *different enough* from the old scoring code, then the scoring code may not need to change at all! If a customer record gets scored into a different segment altogether, then again it's time to review the cluster model in general. The changing of the model should really be a business decision combined with the analysis of the final set of comparative results in order to make the decision when to refit the cluster model. You could profile these scored results and see if the profile looks significantly different from the original clustering segment profile. When I say significant here, I mean *practically* significant. The final process flow diagram is shown in Figure 7.13.

Figure 7.13 Final Process Flow Diagram

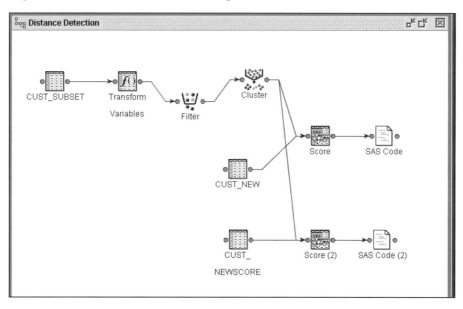

7.4 Other Practical Considerations

One of the many attractive features of creating scoring code from a cluster model is that in very large data sets (e.g., ones with perhaps hundreds of thousands or millions of database records) the sampling of the large data set can be easily obtained using SAS Enterprise Miner (refer to Sample Node in the SAS Enterprise Miner node reference in the online documentation). This can therefore make the clustering model much easier to perform. When the number of observations to cluster gets this large, the time it will take between each pass through the data set can be very long (in numbers of hours). The key to making the sampling work is to ensure that the sample data set is *representative* of the original data set on all the variables that will be used or even the variables that might be under consideration. Then, once the cluster model is developed, the score code can be run against the complete data set to score the clustering model without the excess time it would have taken to cluster the entire data set.

Missing values in your data set can cause some issues for cluster algorithms as do outlier values. In SAS Enterprise Miner, one way to deal with missing data is to set the cluster node missing value property to None. Otherwise, the default causes missing values to be ignored during the training. If the imputation method is set to nearest seed, then the missing value after training will be set to the nearest available seed. If all values for the variables in the cluster model being run are missing, then the entire record is not used in the cluster analysis. Missing values will be discussed in greater detail in Chapter 9, "Clustering and the Issue of Missing Data," when alternatives for ignoring the missing values are discussed.

7.5 Additional Reading

Davies, D. L., and D. W. Bouldin. 1979. "A Cluster Separation Measure." *IEEE Transactions On Pattern Analysis & Machine Intelligence.* 1:224–227.

Halkidi, Maria, Yannis Batistakis, and Michalis Vazirgiannis. 2001. "On Clustering Validation Techniques." *Journal of Intelligent Information Systems.* 17.2/3:107–145.

Using Segments in Predictive Models

8.1 The Basis of Breaking Up the Data Space

One of the most common methods in data mining for understanding patterns in customer purchase data, predicting the value of a customer at some future date, or other issues similar to these, is to make it much easier for the data mining algorithm to *learn* the desired patterns and generalize them rather than *memorizing* the patterns. Generalizing is usually when the algorithm learns enough of the signal pattern in the data that it can be successfully used to score new records never used in the training or validating of the model. When an algorithm memorizes a data set, statisticians typically call this an over-fit model. Over-fitting fits the data so well that when new data records are scored, they are so different from the training set that they don't predict in the fashion that the analyst had intended. This can sometimes be observed during the training of a Neural Network model as the error rates in the training data set get better over time, the validation data set gets worse, and the model results on the this validation set become too far away from the training set. When there are many variables on a data set, dimensionality is a fundamental challenge to create classifications and/or predictions. As the number of dimensions grows, it become increasingly more difficult to construct a fully specified model since the model complexity tends to grow faster than the number of dimensions grows. This is typically referred to as the *curse of dimensionality*. One technique at reducing the level of this curse of dimensionality is to break up the data into different chunks so that a model can learn each chunk better than if the entire set of data is used for training all at once (Hand, Mannila, and Smyth 2001, pp. 187–189). As a simple analogy, review the data plotted in Figure 8.1. Although this data could be represented by a simple polynomial function relating the response variable Y with the single variable X, a simple method using only linear functions in X can be used, which is represented by the dashed lines. This is referred to as a *piecewise linear approximation* of the data.

Figure 8.1 Simple Analogy of Piecewise Linear Data Approximation

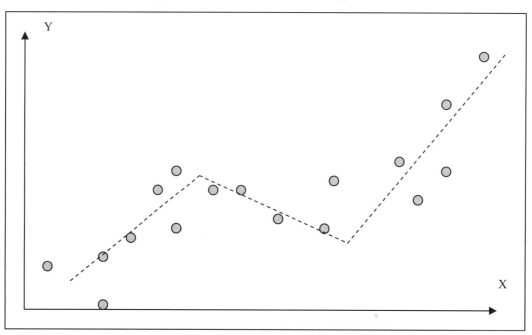

When many variables and dimensions exist, the data space grows rapidly and becomes difficult to model and even visualize. Factorization of a density function into component parts helps supply a general technique for construction of simpler models for multivariate data. For example, if we assume a set of variables is modeled by a density function called p_k and if each function acts independently of each other, then we can write an overall density function as shown in the following equation (Hand, Mannila, and Smyth 2001, pp. 187–189):

$$p(x) = p(x_{1,....}x_2) = \prod_{k=1}^{p} p_k(x_k) \tag{8.1}$$

where x are the data elements, and p represents the one-dimensional density function for the set of x_k elements.

It is usually much simpler to model a one-dimensional density function several times than to model a group of them all together. The independence allows each of the density functions to be multiplied together.

If we apply this concept to segmentation, we can use the technique of clustering to help break up the data space according to a set of desired variables (attributes) and then select a desired segment to create a predictive model. Sometimes, more than one model can be used for a segment, and a final model might be a combination of the portions from the component models. A combination *ensemble model* accomplishes just that. It takes the best set of predictions from each of the component models and often produces a better model than any individual component model is able to perform.

In a business example, certain customers in a database are considered very loyal and valuable. When these customers have been segmented into a group of like customers, we could predict which set of prospects would be very similar to the particular segment that is valuable in the customer base. Sometimes, direct marketers call this a *look-alike* model because the prospects look very similar to the set of desired customers. We will attempt to perform this type of model in this chapter.

8.2 Predicting a Segment Level

Segments are not necessarily predictive; however, they generally are descriptive, and they are a type of classification as we have seen in earlier chapters. In business applications (especially in CRM), it is desirable to make a classification (which is really a certain class of prediction) and predict that class in a set of data where that classification cannot or does not exist. As briefly described in Section 8.1, one good way to find potential prospects is to look in the same place where your best customers are in your database. That means having a method of determining who are your best customers and it also means having a set of criteria in which to test who is a best customer compared to all the other customers. In Chapter 4, "Segmentation Using a Cell-based Approach," and Chapter 5, "Segmentation of Several Attributes with Clustering," we saw that RFM is one technique for classifying customers on their purchasing history. Because customers change their patterns over time since they first became a customer, it is a good idea to track your customers and know what they looked like back when they were a prospect or just after they became a customer (Berry and Linoff 2004, p. 108). A method for knowing what the customer looked like could be obtained using the profiling technique described in Chapter 2, "Why Segment? The Motivation for Segment-based Descriptive Models," and in subsequent chapters. So, let's try to predict a segment level after having performed some segmentation.

Process Flow Table 1: Predicting Segments Project

Step	Process Step Description	Brief Rationale
1	Start SAS Enterprise Miner and create a new Predicting Segments project and process flow diagram Buyer Segmentation; add the BUYTEST data.	
2	Add the BUYTEST data set onto the flow diagram.	
3	Drag a Cluster node and connect the BUYTEST data to it.	Clusters the BUYTEST data into segments.
4	Select specific variables for the cluster segmentation.	Shows the variables that are known to be correlated with each other; only one should be used.
5	Review the cluster segment profiles.	Shows what the clustering found.
6	Add a SAS Code node to create a target variable from cluster 5.	Makes a target from a desired segment.
7	Drag a Metadata node to change the variable target to a target role.	Changes the role of a variable from input to target.

(continued)

Process Flow Table (*continued*)

Step	Process Step Description	Brief Rationale
8	Add a Data Partition node and connect the Metadata node to it.	Sets up train, validation, and test data sets.
9	Select an appropriate model for the analysis.	Considers client and analytic needs.
10	Drag a Regression node and select variables and properties.	Shows results on the regression run.
11	Revise the regression to add interaction effects.	

Step 1: If you have not done so already, open SAS Enterprise Miner and create a new project called Predicting Segments. Now right-click the Data Sources folder icon and select Create Data Source. Then walk through the next several screens selecting a SAS table from the metadata source screen, browse the SAS libraries to select the SAMPSIO library, and select the BUYTEST data set. Create a new process flow diagram and call it Buyer Segmentation. In this flow, we will perform cluster segmentation and then predict a level of one of the segments. **Step 2:** Drag the BUYTEST data source onto your process flow workspace. **Step 3:** Now add a Cluster node and connect the BUYTEST data source to the Cluster node. For this example, we'll use only certain variables and select only one of several that are highly correlated to each other. When using variables for clustering, several variables that are highly correlated to each other could throw off the clustering algorithm in a similar fashion as in a regression. For example, the variables BUY6, BUY12, and BUY18 are all perfectly correlated as the definition of BUY6 is the number of purchases a customer made within 6 months; BUY12 is the number of purchases in 12 months, etc. Therefore, only one of these variables should be used. **Step 4:** In the Cluster node, click the Variables selection and remove the variables from clustering as shown in Figure 8.2. Be sure to use the ID variable as that identifies that each customer is a unique identity. In the Cluster node, ensure that the Internal Standardization is set to Range as the default is None. The other properties can remain at their default settings. Now run the Cluster node path.

Now, you need to profile the segments that the clustering algorithm found. You should have five cluster segments of approximately equal sizes. **Step 5:** Open the results section of the Cluster node. Figures 8.3 and 8.4 show the initial clustering segments and the cluster distances plot. Clusters 2 and 3 are rather close to each other, while clusters 1, 4, and 5 are well separated.

Figure 8.2 Variables to Use in Cluster Segmentation

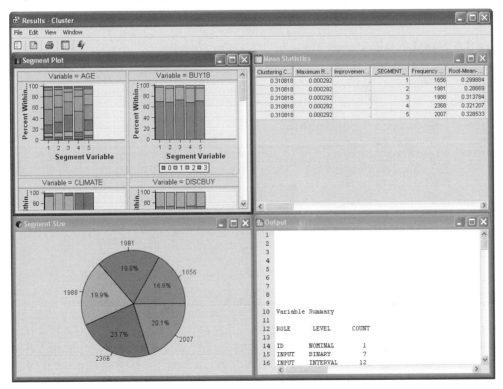

Figure 8.3 Clustering Segment Results Window

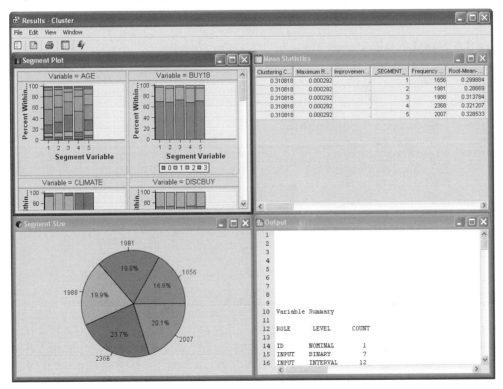

Figure 8.4 Cluster Distances Plot

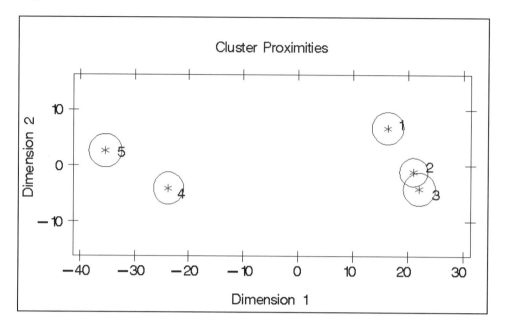

Profiling these segments can be done in a similar fashion to what was shown in Chapters 5 and 6. From the Variable Importance table in the Cluster node results in the **View→Cluster Profile** menu, the top three variables are location of residence, total value of purchase in last 24 months, and income in thousands. These are shown in Figure 8.5. Although clusters 1 and 2 are reasonably close together, the main difference is that in cluster 1, almost all the customers own their own home, and in cluster 2, almost none of the customers own their own home; clusters 3 through 5 have a mix (62%, 13%, 7.5%), respectively.

Figure 8.5 Variable Importance Table from Cluster Results

NAME	LABEL	RELATIVE IMPORTANCE
LOC	Location of residence, A-H	1
VALUE24	Total value of purchases last 24mo	0.864479
INCOME	Yr Income in thous.	0.846406
OWNHOME	1 if own home, 0 otherwise	0.772101
PURCHTOT	Test mailing purchase total by product category	0.742229
CLIMATE	Climate code for residence, 10, 20, 30	0.734116
SEX	F or M	0.649197
AGE	Age in years	0.605777
FICO	Credit Score	0.531327
BUY18	# of purchases 18mo	0.497759
MARRIED	1 if Married, 0 otherwise	0.455787
RETURN24	1 if product return in past 24mo, 0 otherwise	0.056667
DISCBUY	1 if discount buyer, 0 otherwise	0.055788
ORGSRC	Original customer source; C, D, I, O, P, R, U	0

What makes cluster 5 stand out from cluster 4 or other clusters is variable VALUE24 (the total value of purchases in the last 24 months). The average of VALUE24 is the highest in cluster 5 than any of the other clusters; however, cluster 5 does not have the largest average income. For example, cluster 5 has an average income of $48.07 (thousands) and the average total value in 24 months is slightly more than $288. Contrast this with cluster number 1 with an average income of

$56.26 (thousands) and an average total value in 24 months of $236.9. These statistics give us an indication that in this data set the wealthiest customers do not always purchase the most goods.

If a marketer would like to obtain more customers like those in cluster 5, the marketer (and analyst) has a couple of options. On one hand, in order to get more customers like those in cluster 5, the marketer can increase the value of other customer segments and thus grow the customer base. Indeed this is a great tactic as other customers already have purchased in the past and therefore may be more likely to purchase again, so campaigns designed especially for them could be developed. However, if customer acquisition were a priority, then the marketer would like to obtain customers who eventually will look like those in cluster 5. The major assumptions that need to be looked at for developing a predictive model from variables in a customer data set and then scoring those predictions onto a prospecting data set are as follows:

- The data in the prospecting base should be similar enough to the customer data in order to generate high enough predictive scores; this is not a requirement, however, but reasonable practical advice. Reasonable, because if the scores don't reflect the desired attributes then the scoring model won't produce the desired result for the marketer. The developed model should be of good business value as well as technically sound.

- The variables used in the prediction model need to be available in the prospecting data set. This is a requirement, however, if one or more variables in the prospect data set are not there but are in the model, then the model will not perform as intended. In fact, it will generally perform poorly depending on the algorithm used to build the model.

- The levels of categorical variables and distributions of the numeric ones on the customer data need to be somewhat representative of the data in the prospecting data set.

- The variables used to predict the cluster segment(s) should be available in the prospecting database.

So now that we think that cluster 5 should be a good level to predict, for our example let's assume that all the variables on the prospecting data set are available on the customer data set that we will be using to develop our predictive model. **Step 6:** Drag a SAS Code node onto your process flow workspace and connect the Cluster node to it. Open the SAS Code window by clicking the SAS Code icon in the property sheet. Enter the SAS code as shown in Figure 8.6.

Figure 8.6 SAS Code to Generate the Target Variable

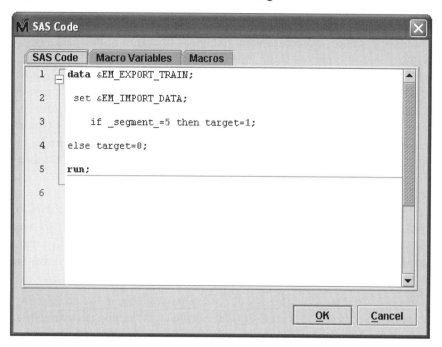

The code will create a new variable called TARGET, which will be binary and equal 1 when the cluster segment variable _SEGMENT_ equals 5; and 0 when all other cluster segment levels equal 0. This new TARGET variable will be our response we want to predict. If you want all 5 levels to be predicted, then you don't need to generate a predictive model at all. SAS Enterprise Miner will create scoring code from the Cluster node as demonstrated in Chapter 7, "When and How to Update Cluster Segments." The automatic SAS macro variable &EM_EXPORT_TRAIN will automatically contain the SAS data set for the newly created training data and the output from the Cluster node is captured in the SAS macro variable &EM_IMPORT_DATA. **Step 7:** The next step is to change the TARGET variable into a response instead of an input, so drag a Metadata node onto the flow diagram and connect the SAS Code node to it. Then right-click and select the Update ⟳ icon. This will update the Metadata node with the newly created data set from the SAS Code node. Now open the Metadata node variables icon in the property sheet and set the new role for the TARGET variable to the Target role as shown in Figure 8.7. This will indicate to SAS Enterprise Miner that the TARGET variable is a variable we want to predict in a model. Also, set the TARGET variable to Binary instead of the default value.

Figure 8.7 Changing the Role of the TARGET Variable

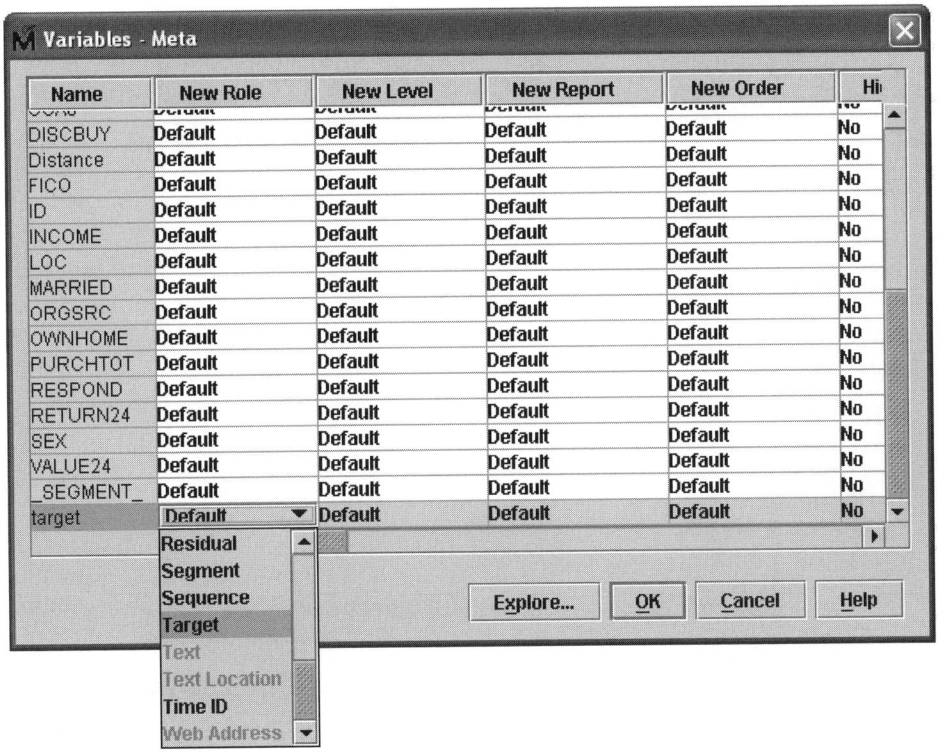

Step 8: Now we need to set up the Training, Validation, and Test data sets so drag a Data Partition node and connect the Metadata node to it. In the property sheet of the Data Partition node, set the Partitioning Method to Stratified and ensure that the TARGET variable indicates Stratified. This will ensure that each of the three data sets (Training, Validation, and Test) will have the same proportion of the TARGET variable as in the original data set. It will prevent one of the three partition data sets from having no response levels at all or the wrong proportions.

Step 9: We now have a choice to make as to which type of model to select. We could choose one or perhaps we could try several models and compare them to each other. In business situations, it is important to take note of your client's needs. For example, if your client needs a predictive model for the prospecting scores but does not really care as to the level of explicability, then as an analyst you could choose a Neural Network model and not have to worry about explaining how the model works or what variables were most important in the model, etc. However, if a client does want these attributes, then it's best to have a model in which a profile and description of the model can be given in relatively simple business terms. A regression or decision tree model can be easily explained. For simplicity, let's choose a regression and since our response variable (TARGET) is binary, we'll be fitting what is called a *logistic regression*. Logistic regression predicts the probability of a binary or ordinal variable (e.g., probability of the value being 0 or 1, yes or no, or perhaps male versus female) as examples of binary variables. Logistic regression can also be used to predict a categorical variable with four or even five levels; however, as the levels increase, the difficulty in obtaining a well fitting model also increases considerably!

Step 10: So now drag a Regression node (not the one called Dmine Regression[1]) onto your process flow diagram and connect the Data Partition node to it. Your process flow diagram should now look like the one in Figure 8.8.

---

[1] In SAS Enterprise Miner 5.2 there are two regression nodes: Dmine and Regression.

Figure 8.8 Cluster Segmentation and Predictive Model with Regression

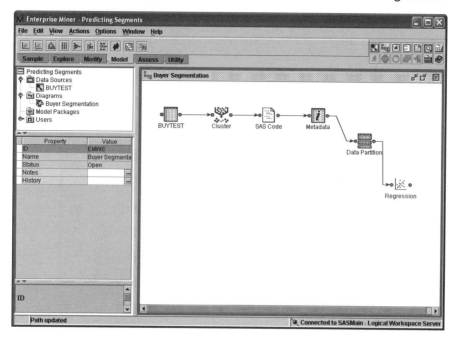

In the Regression node property sheet, I typically set the Input Coding to GLM rather than the default level of Deviation. This is a matter of preference rather than a technical choice in this case. The GLM coding compares all levels to the sorted last level of each variable. The Deviation coding compares each level. Figure 8.9 shows the variables we'll initially try in our regression model run. You could at this point select Forward or Backward in the Optimizations options; however, I like to see the results of the model as I select the variables for the model. This is mainly due to practical considerations of using variables that are important to the business rather than a statistical procedure selecting the variables. Both of these techniques can be used here, but for this simple example, we'll select the variable and see the results and then perhaps alter the model accordingly.

Figure 8.9 Variables Initially Selected for the Logistic Regression of the Target Variable

Name	Use ▲	Report	Role
RETURN24	Default	No	Input
ORGSRC	Default	No	Input
OWNHOME	Default	No	Input
DISCBUY	Default	No	Input
LOC	Default	No	Input
SEX	Default	No	Input
PURCHTOT	Default	No	Input
MARRIED	Default	No	Input
VALUE24	Default	No	Input
RESPOND	Default	No	Input
Distance	Default	No	Rejected
INCOME	Default	No	Input
COA6	Default	No	Input
CLIMATE	Default	No	Input
AGE	Default	No	Input
BUY12	Default	No	Input
FICO	Default	No	Input

Now you can close the Variables window once you have selected these variables and run the Regression node. Run the Regression node and then open the Results window once it completes. At this point, what we have done so far is to perform a clustering segmentation and run a first pass at a regression to predict cluster number 5. Figure 8.10 shows the process flow diagram to this point and your initial Regression Results output window is in Figure 8.11.

Figure 8.10 Process Flow Diagram for Cluster Segmentation and Initial Regression to Predict Segment Number 5

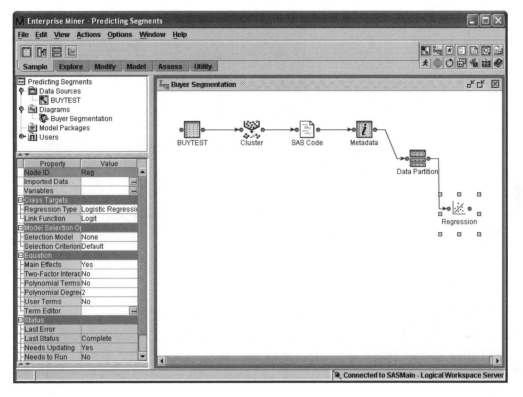

In Figure 8.11, the default settings for a binary target response show the Score Rankings as a lift chart. The vertical axis displays the cumulative lift, and the horizontal axis displays the decile of the training data set. By default, the blue curve is the Training data set and the red curve is the Validation data set. The two curves should be somewhat similar if the statistical sampling performed in the Data Partition node was sufficient and the regression does not contain problems or issues regarding variables that are highly correlated to each other. The lift chart indicates how much more the model is predicting the response target versus from just selecting the data at random. High values of lift over a long segment of deciles are desired.

The Effects Plot bar chart indicates the magnitude (in order of largest to smallest) of each variable's contribution for predicting the target levels. You can place your cursor over each effect bar to see its relative magnitude and the name of the variable. The Output window shows the output of the regression statistics from fitting results (also in the Fit Statistics window) and other parts of the regression. One of the more useful parts is assessing the model fit and the Type 3 analysis of effects table. This table indicates the statistical significance of each variable in the model with respect to the target responses. The larger the chi-square value, the smaller the p-value, and thus the greater the statistical significance for that variable. Our current model is a *main effects* model in that we don't have any second order effects such as interactions between variables, only the main effect of each variable. The typical value for significance is to look for p-values lower than 0.05; however, even variables that have a good business importance for being in the model with a p-value of 0.2 could still be a valid selection.

Figure 8.11 Initial Regression Results Window

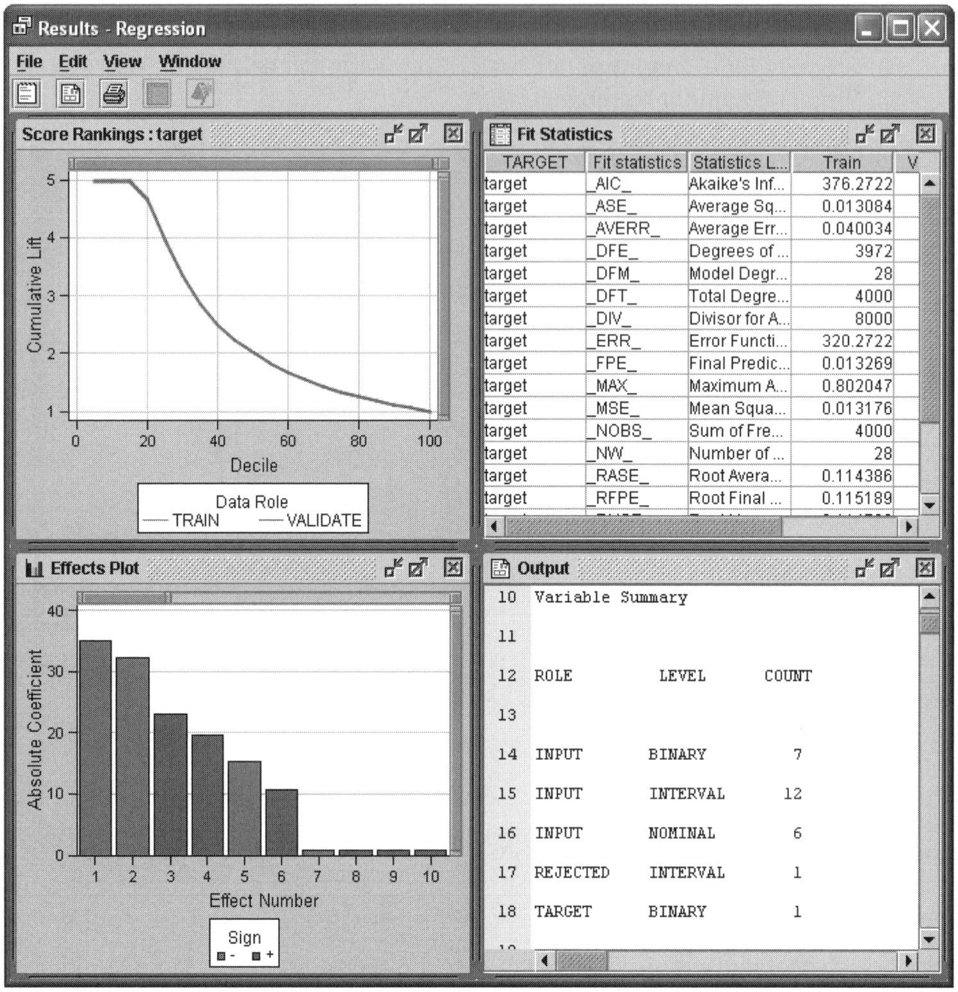

In our first run for this model, the effects that stand out are CLIMATE, SEX, MARRIED, and OWNHOME. Let's now try to add in a second order effect. **Step 11:** Click the Regression node, then in the Properties sheet click the User Terms, and highlight Yes. Now click the Term Editor. Select the variables MARRIED and OWNHOME, click the Save button, and then select OK. This will add the interaction of these two variables to the main effects model. Now, you should rerun the Regression node. When you open the Results window and look into the Output, you should see the Type 3 Effects table after scrolling down a bit. This is shown in Figure 8.12. Notice that the main effects p-values are below .001 and the p-value of the interaction effect between the two denoted as MARRIED*OWNHOME 0.0528. You can place your cursor over the Effect Plot and see the result of each effect (blue being negative and red being positive) that affects the outcome variable called TARGET. To aid in our assessment of this model, drag onto the workspace a Model Comparison node and connect the Regression node to it. Now run the Model Comparison node. In the results of the Model Comparison node, several Lift and ROC curves are plotted as shown in Figure 8.13.

Figure 8.12 Output of Regression with Two-Factor Interaction Added

```
Output                                                      ⌐ ⌐ ⌐ ⌐ ⊠
150              Type 3 Analysis of Effects

151

152                              Wald

153  Effect           DF     Chi-Square     Pr > ChiSq

154

155  AGE               1        0.0076        0.9304

156  CLIMATE           0          .             .

157  DISCBUY           1        0.0020        0.9641

158  LOC               5        0.0085        1.0000

159  MARRIED           1       17.1170        <.0001

160  ORGSRC            6        0.0107        1.0000

161  OWNHOME           1       19.5918        <.0001

162  RESPOND           1        0.0000        0.9956

163  RETURN24          1        0.0000        0.9954

164  SEX               1       30.1304        <.0001

165  VALUE24           1        0.0029        0.9572

166  MARRIED*OWNHOME   1        3.7510        0.0528
```

Figure 8.13 Model Comparison Node Lift and ROC Charts

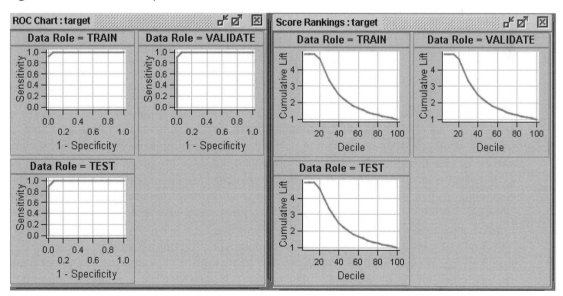

From the lift charts on the Training, Validation, and Test data sets, at the 50th decile points, the cumulative lift is about 2X and above. This means that the model is capturing the actual responses at twice the rate as compared to the rate if no model were used at all (e.g., randomly selecting the population audience). In the first two deciles, the amount of lift is 5X and above. By right-clicking the charts, you can select Data Options, which allows you to change the metrics plotted on the X and Y axes and also allows you to modify the appearance of the graphics. We have not predicted cluster 5 from the previous cluster segmentation, and now we can use it to score other data set records.

8.3 Using the Segment Level Predictions for Customer Scoring

At this point, we have created a clustering model and a predictive model that predicts one cluster segment level obtained from the clustering algorithm. How can this exercise be useful in a business context? We believe that cluster 5 in the previous exercise does indeed represent the more valuable customers, however, not the customers with the largest income. So when predicting cluster 5, we would be trying to find customers with approximately the same income level, but they would be more likely to purchase in a similar fashion to the customers found in cluster 5, according to our model that is. This type of predictive model comes in handy with a prospect database, which had the same variables as the customers except for purchase fields (like VALUE24). So, to be clear, if cluster 5 is to be predicted and scored on a *prospecting database*, then you can use *only* the fields on the *prospecting database* for predicting this segment 5! If this is successful, then a model could be developed using only the variables available in the prospect database and you would now have a method for scoring the probabilities of the *prospect database* for cluster 5. You could also predict cluster 1, which had the highest income, and this might represent an opportunity for customer growth if you can find what product portfolio is best suited for cluster 1. We will address this issue in greater detail in Chapter 10, "Product Affinity and Clustering of Product Affinities."

8.4 Creating Customer Value Segments

Knowing which customers are valuable, which customers lack potential, and which customers should be *grown* to develop their value is vital for business profitability. Customers that are clearly more profitable should actually be treated differently from customers who are not. This does not mean you should *mistreat* any of your customers. With all due respect, each and every customer deserves courteous, prompt, and fair attention. What I am indicating here is certain customers should have *different* types of attention than others. For example, if you have signed up for a frequent flyer program with an airline provider and you use this airline almost exclusively, then you have given them enough business to warrant something extra for being such a good customer. This might be some added bonus points, exclusive entry into their executive lounge club, or whatever program, campaign, or special offer they might have for their *best* customers. If this is the case then, before anyone can offer these kinds of programs to the customer, the airline provider must have a method for determining which set of customers qualify for those programs and which do not. This is where understanding and creating customer *value segments* comes into practice.

In Chapter 4, we reviewed segmentation from an RFM standpoint, and the airline or whatever company could use that RFM classification scheme to offer the highest RFM categorization for the determination of the *most valuable customers*. However, RFM is really only a grouping of three attributes and two of them (recency and frequency) are somewhat related to one another. What if the business would like to determine its *most valuable* customers based on perhaps more attributes than just two or three, such as how long that customer is expected to stay a *best customer*. On the other hand, instead of net revenues, maybe profit margin for each customer is computed and used in its place. The customer's potential (both monetary and needs based) is an additional attribute that might play into the criteria for a *valuable customer*. The business should determine what it considers a valuable customer and what it does not. The data mining analyst should know who could best determine what combination of attributes the business needs for its *most valuable* customer segments.

Economists have typically indicated that in order for a company to increase value for the shareholders, business decisions should seek to increase this value and all projects should have some impact in this increase in value. Marketers will do well to guide their efforts at maximizing this value as well. Customers usually vary widely in their value to a business due to differing spending patterns, loyalty, needs, and their propensity to spawn referrals (Rud 2001, p. 283). Segmentation should include the considerations of the lifetime value (LTV) of customers. There are a number of methods for determining the LTV of a customer. Computing LTV for products is somewhat different from a renewable service such as a service contract. You could combine an estimated probability that a customer will stay a customer in some future time period with value and hazard models that perform that function (Potts 2005). *Hazard modeling* is a little beyond the general scope of this book; however, there are other computation methods for determining the lifetime value of a customer.

Increasing the customer LTV using a customer metric is very attractive for the following reasons:

- Increasing the LTV of customers increases the value of the company.

- Customer LTV can be directly tied to important marketing and sales objectives such as targets and retention or loyalty.

- Calculations for LTV require the marketer to view the customer in the long run and more comprehensively.

- Accounting for LTV differences is directly related to the differing levels of risk and customer profitability (Rud 2001, p. 283, and Peppers and Rogers 2005).

Lifetime value computations have several components depending on the type of business and whether the goods and services are renewable. First, you have *duration*. Duration is the expected length of the customer relationship. By relationship, I mean purchase relationship rather than relationship to a sales person. The purchase relationship is what will matter most when considering the value to the shareholder or the owner of the business as in a privately owned business. As mentioned earlier, duration or probability of continuing the relationship can take one of many forms depending on your level of sophistication in these types of estimates. *Time period* is the length that LTV will take for the desired increment. This is typically a year, but other renewal periods or product cycles could be used. *Revenue* is the income from the sale of goods (product) or services. *Discount rate* is the adjustment to convert a future dollar value into today's value. This is incorporating the time value of money. *Costs* are the marketing expense or direct cost of the product or service. The cost associated to customers does not typically include the cost of buildings, facilities or other costs. The cost we are after here needs to be *directly* related to the customer. You should check with your finance department (or other knowledgeable workers) as they might be able to help you to better define how costs are allocated within your business. *Renewal rate* is the probability of renewal or retention rate. Again, the retention rate could be determined with a predictive model or perhaps computed directly depending on the business need and type of business complexity. *Risk factor* can be incorporated to include the potential risk related to losses such as a customer returning items, not paying by going bankrupt, etc.

The business case for this example is as follows. A customer-based marketing manager would like to send out a communication of a special event where the VP of sales is a featured speaker, and the communication needs to go out to all of the *most valuable* set of customers. These are not the customers who necessarily generate the most revenues, but the ones that generate the most *profit* and do so over the long-haul. This means that we would need the three-year LTV estimate on each customer. Obviously, there are many things one can do with the information that a particular set of customers are considered *most valuable,* however, this example will suffice for the moment.

Process Flow Table 2: Most Valuable Customers (MVCs)

Step	Process Step Description	Brief Rationale
1	Start SAS Enterprise Miner and create a new project MVC Segment and a new diagram MVC Process Flow.	
2	Add the data CUSTOMER_VALUE to the data sources folder.	
3	Data Partition node to split off training, validation, and test data sets.	Ensures that the LTV model works on a holdout sample (test data set).
4	Transform Variables node and steps to transform certain variables.	Makes the variables appropriate for modeling.
5	Metadata node to review variables and set REV_THISYR to target.	Sets up REV_THISYR to predict.
6	Regression node to predict the target variable.	
7	Add a StatExplore node for Data Assay analysis.	Reviews the REV_THISYR variable.
8	Revise the Regression node with variables that are not significant.	Uses only the variables that contribute to the regression (explaining variance the most).
9	Add a Score node to the diagram and another CUSTOMER_VALUE data set (role of score).	
10	Drag a SAS Code node and add SAS statements for LTV.	Computes final calculations.
11	Enter SAS code to compute the LTV estimates.	Computes final calculations.

Step 1: In this example, we will look at a set of profitable customers and a method for determining the most valuable set of customers as well. Now let's create a new project and call it MVC Segment (for most valuable customer segment) and a new diagram called MVC Process Flow. **Step 2:** Select **Data Sources→Create Data Source**. Click SAS Table and select the Browse button to view the available SAS data libraries. In the SAMPSIO library, select the data set CUSTOMER_VALUE. Continue through the Data Advisor options, click the Advanced button, and click Next. A window will show you the columns and their basic attributes. You have the opportunity in this window to change the attributes as SAS used the advanced data advisor settings to guess the type for each variable. If the variable STATE is set to rejected, then select the role as Input and ensure that the variable level is set to Nominal. Click the Next button and now this data set is added to the data sources icon. You should now be able to drag the CUSTOMER_VALUE data set (actually the metadata, not the real data) onto your diagram workspace.

Figure 8.14 Customer Value Data Set Variables in the MVC Example

Name	Role	Level	Report	Label
CITY	Rejected	Nominal	No	
PURCHFST	Input	Nominal	No	Year of 1st Purchase
PURCHLST	Input	Nominal	No	Last Yr of Purchase
RFM	Input	Nominal	No	Recency, Freq, & Monetary Value Code
SEG	Input	Nominal	No	Industry Segment Code
STATE	Input	Nominal	No	
channel	Input	Nominal	No	Purchase Sales Channel
corp_rev	Input	Interval	No	Corporate Revenue last fiscal yr.
cost	Cost	Interval	No	
cust_id	ID	Nominal	No	Customer ID No.
customer	Input	Nominal	No	A=New Acquisition, C=Churn (no purch), R=Cont-Purchase
est_spend	Input	Interval	No	Estimated Product-Service Spend
loc_employee	Input	Interval	No	No of local employees
profit	Input	Interval	No	
public_sector	Input	Binary	No	0-No, 1=Yes
rev_class	Input	Nominal	No	Revenue Class Code
rev_lastyr	Input	Interval	No	Last Years Fiscal Revenue
rev_thisyr	Input	Interval	No	This Years Fiscal Revenue YTD
tot_revenue	Input	Interval	No	Revenue for All Years
us_region	Input	Nominal	No	US Region Location of Business
yrs_purchase	Input	Nominal	No	No of Yrs Purchase

Figure 8.14 shows the variables and labels used in this example. To open this window, click the Variables selection in the Input Data node property sheet. Notice in the listing of variables that a cost variable is listed. This data set is similar to the CUSTOMERS data set used back in Chapter 5 and its subset in Chapter 7. The costs associated with products and marketing are contained in this COST variable for each customer record. We will use this to help determine the most profitable set of customers by computing an estimate of profit for each customer record and their LTV for the next year. We need to estimate one year of net revenues beyond this year and then project revenue for three years' time based on past revenues. **Step 3:** Be sure to drag onto your workspace flow diagram a Data Partition node and set the three-way splits to 65% for training, 15% for validation, and 20% for the test data sets. Forecasting revenues, however, requires a set of transformations so the steps we will take to compute the net revenues for the next three years is as follows: **Step 4:** Add a Transform Variables node and connect the Data Partition node to it.

- Transform last year's net revenues and this year's net revenues so that they look relatively *normal* as in a normal distribution. Predicting a distribution of the original net revenues is rather difficult. To see this, in the CUSTOMER_VALUE node use the Explore button to view the distributions of variables REV_LASTYR and REV_THISYR.

- Next, we will set the transformed variable REV_THISYR as our predictor variable. **Step 5:** To do this, drag a Metadata node and connect the Transform node to it. Click the Variables icon, or right-click the Metadata node and select the Edit Variables option. You can explore the transformed revenue distributions once the previous portion of the diagram has been run. Figure 8.15 shows the variable LOG_REV_THISYR variable distribution. Next, set the variable LOG_REVTHISYR as the target variable in the New role field.

- We will create a predictive model to estimate the transformed REV_THISYR.

- **Step 6:** Now, drag a Regression node onto the workspace and connect the Metadata node to it. In the Regression node's property sheet, we'll want to first assess the main effects. That is, we'll only look at each variable's single contribution in the model. Then, perhaps we'll look at some interaction and/or non-linear effects. Figure 8.16 shows the property sheet selections for the Regression node.

- Before we run the Regression node, we might first want to review whether the variables in this data set are somewhat correlated to each other. This will greatly affect the regression results if you have two or more variables that are highly correlated to each other in the same model. **Step 7:** You can do this by attaching a StatExplore node and ensuring that the Correlation fields are set to Yes. Now, after you run the StatExplore node, open the results to see correlation statistics. Any field that is larger than approximately 0.4 for the Spearman's or Pearson's correlation coefficients should be considered "correlated." From the results of this analysis, PROFIT and LOG_TOT_REVENUE are both above the 0.4 level. It would be best for only one of those variables rather than both to be placed in the model since their correlation is strong enough that it could cause the regression model some problems.

- Now you can run the Regression node. Once complete, open the Results browser and view the Output window to review the printed regression output results. Figure 8.17 shows the Type 3 statistics for the variables in the regression obtained from the Output window. These statistics indicate how strong or weak each variable's contribution is to the target of the regression model. A Pr > F (read p-value greater than the F statistic) smaller than .05 indicates that the variable is statistically significant at the 5% level; this means that you have about a 95% chance that this variable is affecting the target variable and thus a 5% probability that this happened by pure chance.

- After reviewing this output, let's take a second pass at a regression model by removing factors (variables) that seem to have little effect on our target variable. **Step 8:** So, edit the variables and remove from the regression's analysis any variable that has a p-value greater than 0.25. This would indicate the removal of STATE, LOC_EMPLOYEE, PUBLIC_SECTOR, and US_REGION. After you have done this, rerun the regression node and check the model fit statistics. Notice now that the R-squared has increased while removing several variables.

Figure 8.15 Distribution of the Transformed Variable LOG_REV_THISYR

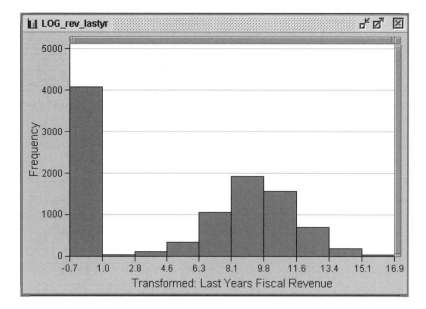

Figure 8.16 Regression Node Property Sheet Settings

Property	Value
Variables	...
Equation	
Main Effects	Yes
Two-Factor Interactions	No
Polynomial Terms	No
Polynomial Degree	2
User Terms	No
Term Editor	...
Class Targets	
Regression Type	Linear Reg
Link Function	Probit
Model Options	
Suppress Intercept	No
Input Coding	GLM
Min Resource Use	Default
Model Selection Options	
Selection Model	None
Selection Criterion	Default
Selection Default	Yes
Sequential Order	No
Entry Significance Level	0.05
Stay Significance Level	0.05

- Figure 8.18 shows the revised second pass of the regression model with non-significant variables removed.

- **Step 9:** Now, drag a Score node and connect the output of the Regression node to it along with another Input Data source node for the CUSTOMER_VALUE data set. Set the role of this Input Data source to be Score.

- **Step 10:** You should now drag a SAS Code node to your diagram workspace and connect the Score node to it. I've labeled this node Customer Value Calculations.

Figure 8.17 Type-3 Statistics from Regression Model (First Pass)

```
Output                                                          ⌐ᵏ ⌐ᵈ  ⊠
75
76              Type 3 Analysis of Effects
77
78                           Sum of
79  Effect            DF     Squares    F Value    Pr > F
80
81  LOG_rev_lastyr     1     95.3838      69.72    <.0001
82  PURCHFST          16     62.7564       2.87    0.0001
83  PURCHLST          17    100096.474  4303.90    <.0001
84  RFM                7     66.8638       6.98    <.0001
85  SEG               18     54.0106       2.19    0.0025
86  STATE             47     70.0446       1.09    0.3127
87  channel            3     25.8238       6.29    0.0003
88  est_spend          1     14.2875      10.44    0.0012
89  loc_employee       1      0.0384       0.03    0.8669
90  public_sector      1      0.1503       0.11    0.7403
91  rev_class          5    1193.9433    174.54    <.0001
92  us_region          2      0.1452       0.05    0.9483
93  yrs_purchase      17     82.6359       3.55    <.0001
```

Figure 8.18 Type-3 Statistics from Regression Model (Second Pass)

```
Output
76              Type 3 Analysis of Effects
77
78                           Sum of
79  Effect            DF     Squares    F Value    Pr > F
80
81  LOG_rev_lastyr     1     98.0570      71.69    <.0001
82  PURCHFST          16     60.3296       2.76    0.0002
83  PURCHLST          17    100909.458  4339.87    <.0001
84  RFM                7     68.6078       7.17    <.0001
85  SEG               18     65.7679       2.67    0.0001
86  channel            3     24.5861       5.99    0.0004
87  est_spend          1     22.4493      16.41    <.0001
88  rev_class          5    1198.9537    175.32    <.0001
89  yrs_purchase      17     83.4194       3.59    <.0001
```

Step 11: Connect the Input Data from the CUSTOMER_VALUE data to the SAS Code node you've just added. I've renamed the SAS Code node to LTV Computations by right-clicking the SAS Code node and selecting the Rename menu option. You can rename the node to customize your flow diagram nodes as desired. Open the SAS Code node and place the following code in the SAS Code tab as shown in Figure 8.19.

Figure 8.19 SAS Code to Compute LTV and NPV from Predicted Revenues

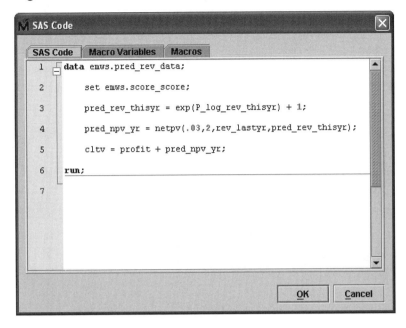

The code in Figure 8.19 takes the scored data set, converts the transformed predicted revenues of this year and the actual revenues of last year, and computes the predicted Net Present Value using the SAS function NETPV. I assumed an inflation rate of 3%, and computed next year's net present value based on the predicted this year's revenues. Then, the customer lifetime value (LTV[2]) is the expected profit added to the predicted NPV. The data set in your workspace library EMWS.PRED_REV_DATA now contains the completed calculations for LTV. Normally, when performing a time forecast, data should be structured as a time series. You can perform a much better time forecast of revenues instead of the method we have here. However, in this example, we had data of only two years to work with and thus a time series would not be feasible with such data. Yet, we can accomplish a fair amount even with such limitations. Figure 8.20 shows the completed process flow diagram of this example.

As I hope you can see, if we used the LTV values we just calculated in the clustering of customers, we might have a slightly different set of clusters than before, and clusters based on the LTV would be showing which customers are more profitable in a future value by using the NETPV function. You should try this as an exercise yourself. Cluster the customers as done in Chapter 5, except use LTV as the primary revenue and don't use the others. Compare and contrast the clusters you obtain with the set of five you obtained earlier.

---

[2] LTV is sometimes referred to as CLTV, denoting customer lifetime value.

Figure 8.20 Completed Process Flow Diagram of MVC Analysis

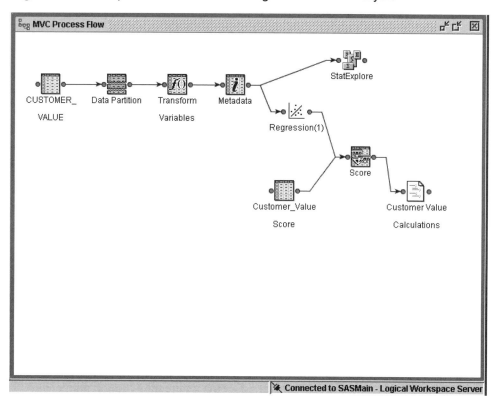

8.5 References

Berry, Michael J. A., and Gordon S. Linoff. 2004. *Data Mining Techniques.* 2d ed. New York: John Wiley & Sons, Inc.

Hand, David J., Heikki Mannila, and Padhraic Smyth. 2001. Principles of Data Mining. Cambridge, MA: MIT Press.

Peppers, Don, and Martha Rogers. 2005. *Return on Customer: Creating Maximum Value From Your Scarcest Resource.* New York: Random House, Inc.

Potts, Will. 2005. "Predicting Customer Value." *Proceedings of the Thirtieth Annual SAS Users Group International Conference.* Cary, NC: SAS Institute Inc. Paper no. 073-30.

Rud, Olivia Parr. 2001. *Data Mining Cookbook: Modeling Data for Marketing, Risk, and Customer Relationship Management.* New York: John Wiley & Sons, Inc.

SAS Enterprise Miner, Release 5.2. Cary, NC: SAS Institute Inc.

Part 3

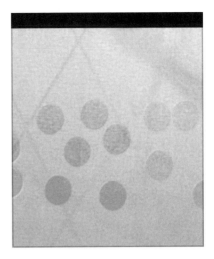

Beyond Traditional Segmentation

Clustering and the Issue
of Missing Data

9.1 Missing Data and How It Can Affect Clustering

When you begin to analyze data, it usually doesn't take too long before you run into the problem of missing data. In just about all typical data sets there are usually missing values for at least one or several variables. In the world of CRM, either in business-to-business or business-to-consumer, missing data can be a real quality problem in any data warehouse. The typical default in most statistical and data mining software applications is if there is a row or record of data that has a missing value for a variable and you use that variable in a regression, for example, in most cases that record is simply omitted in the analysis. The result is a data set input to an analysis engine that contains no missing cases for any of the variables selected for that analysis. Sometimes the literature refers this as *complete case analysis.* For several reasons, this strategy can be both acceptable and non-acceptable, depending on the application of the analytical technique. However, it can also depend on how the analysis is applied and the amount of missing data in the data set. In this chapter, I will review some of the basic principles behind missing data, some tactics for analyzing the patterns of missing data in your data set, and some techniques for imputing missing data both in SAS Enterprise Miner and in a relatively new procedure in

SAS/STAT called the MI procedure for multiple imputation. The effects of missing data will be analyzed with regard to clustering and to segmentation in general since that is what this book is really about. You should keep in mind, however, that many analytical techniques will treat missing data differently. I will briefly point those out, but I won't go into each data mining algorithm and how it's affected by missing data in any great detail.

In Chapter 3 "Distance: The Basic Measures of Similarity and Association," distance metrics were discussed in clustering. Imagine measuring a database record for revenue, for example, and when the distance to another database record is computed, a missing value is in its place. Obviously, you can't measure the distance from one point to another point unless there is a point of measurement. Therefore, that distance metric would be omitted from the distance matrix used to cluster records in the data set. There are, however, some potential alternatives, which SAS Enterprise Miner has built in the Clustering node. We will look at these later. If your data set has 45% missing values for an age variable, then perhaps you don't want to trust the analysis entirely on the 55% of the remaining data alone. If your data set was originally sampled from a much larger set, then when you consider the missing values, the representation may not exist as you intended.

For example, let's say that you collected data on a survey that was conducted on your active customers with a sample of 1,000 surveys, each containing 20 questions. Now, let us assume that this was a telemarketing survey, and not every customer answered all questions; 95% of the customers answered all 20 questions. If the chances of the data missing for one question are completely independent of any other question, then you could expect data on about 360 of the 1,000 surveys in which you have complete answers to all 20 questions! Perhaps you still think that this is fine; however, if you spent $75 per call for the survey and the total amount you spent was $75,000, you would have wasted 48,000 on incomplete surveys. Perhaps, there must be some way to salvage something from all those other survey questions that were thrown out just because one or two questions were not answered on that survey.

9.2 Analysis of Missing Data Patterns

Before attempting any filling of missing data, the first thing you should do is to analyze the amount of missing data in conjunction with the Data Assay introduced in Chapter 2 "Why Segment? The Motivation for Segment-based Descriptive Models." In SAS Enterprise Miner, the StatExplore node gives you some indication of the amount of missing data of your variables in the Output Results window; however, it will do so on each variable independent of any other variables. Another method for evaluating the level of missing data is to classify by groups the unique set of patterns of missing data on the variables of interest. This method then forms a matrix of unique variables and the levels of missing data, which forms a *missing data pattern*. This pattern can take many forms and the monotonic form means that the variables of interest have increasing amounts of missing data elements ordered from left to right. An example of a monotonic missing data pattern is shown in Figure 9.1. Monotonically missing data patterns have some particular properties, which allow certain types of algorithms to be applied when imputing missing data.

Figure 9.1 Example of Monotonic Missing Data Pattern
X = missing, 0 = non-missing data

Pattern	Age	Gender	Weight	# of Subjects in Group
A	0	0	0	1795
B	0	0	X	348
C	0	X	X	475
D	0	X	X	237

The *complete case analysis* referred to earlier has both some attractive features and some major disadvantages as well. The obvious main disadvantage is that in *complete case analysis,* the non-usage of data records due to some variables having missing elements could exclude a large fraction of the original data set. As in the survey case just mentioned, a 5% level of missing cases in a variable may not seem too bad at the outset; however, if a combination of variables all have about that level of missing entries, then that combination of variables will severely limit the analysis and therefore the application of the analysis. There have been many alternatives devised to fill in the missing values; unfortunately, most of these methods are not as good as just using only the complete cases. In recent years, statisticians have developed two methods for handling missing data—*maximum likelihood* and *multiple imputation*—that offer substantial improvements over *complete case analysis* (Allison 2002, p. 2). Although the actual algorithms have existed since the early 1980s, it has only recently become computationally practical within the last several years. In the mid-1980s, personal computers could not even handle maximum likelihood as they were so new to the market and their memory/compute power was severely limited. Even now with computing power as it currently stands, the amount of investment in computing maximum likelihood and multiple imputation can muster a substantial investment of time and energy for learning the methods and in performing them on a regular basis. However, if one desires to have good results, the effort needs to be put forth as the old adage goes, "garbage in, garbage out." Let's review an example of a missing data pattern.

Process Flow Table 1: Clustering with Missing Data

Step	Process Step Description	Brief Rationale
1	Start SAS Enterprise Miner and creating a new project Clustering with Missing Data; create the diagram A Missing Data Pattern.	
2	Add three data sets to your Data Sources folder.	
3	Connect a SAS Code node to the CUSTOMERS data set.	Contains the code to run PROC MI on CUSTOMERS data set.

Step 1: Create a new SAS Enterprise Miner project called Clustering with Missing Data. We will use this project throughout this chapter. Create a new process flow diagram called A Missing Data Pattern. **Step 2:** Add the following data sets into your Data Sources: CUSTOMERS, NYTOWNS, NYTOWNS_WMISSING, and NYTOWNS_IMPUTED, all from the SAMPSIO library. Drag onto your newly created diagram the CUSTOMERS data set and a SAS Code node. **Step 3:** Connect the CUSTOMERS Input Data node to the SAS Code node and place the code depicted in Figure 9.2 into the SAS Code tab of the SAS Code node. Change lines 2 and 3 to reflect a folder of your choosing on your computer; otherwise, you will receive an error message when you run the SAS code. PROC MI is a relatively new SAS/STAT procedure that performs multiple imputation; however, I've put in the PROC MI options statement NIMPUTE=0, which will cause PROC MI to perform only the missing pattern analysis and no imputations or estimations take place. Now, go ahead and run the SAS Code node. The output will still be placed in the Output tab of the results in the SAS Code node; however, the output in HTML format will be placed in the file called MISSING_PATTERNS.HTM in the folder location you've modified. After you run the SAS Code node, open the file MISSING_PATTERNS.HTM with your Web browser; the table of interest is the Missing Data Patterns table. This is shown in Figure 9.3.

Figure 9.2 SAS Code Node to Run PROC MI

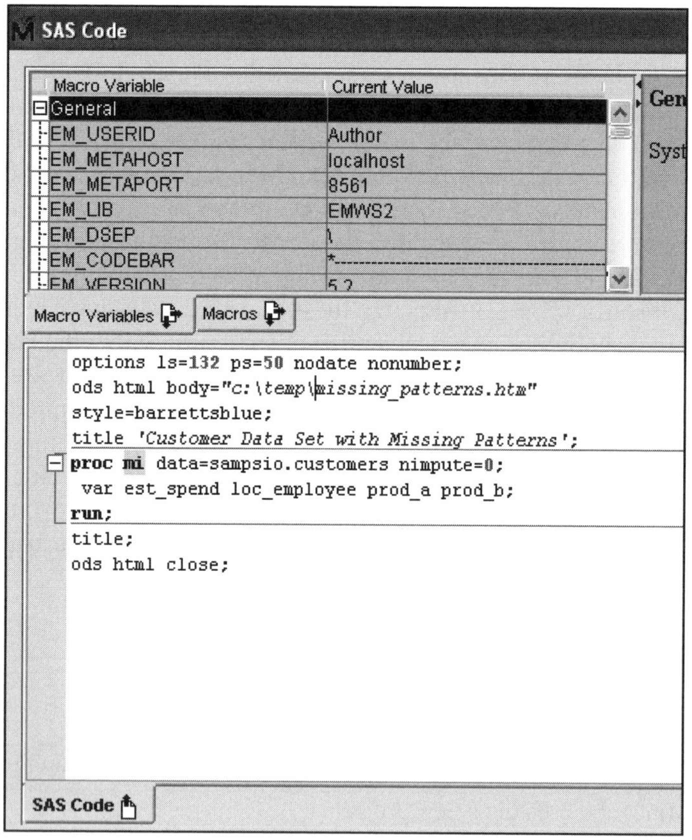

Figure 9.3 Missing Data Pattern Output from SAS Code in Figure 9.2 in HTML Format

Group	est_spend	loc_employee	Prod_A	Prod_B	Freq	Percent	Group Means est_spend	loc_employee	Prod_A	Prod_B
1	X	X	X	X	9006	8.54	746221	401.304908	13.848212	229.915723
2	X	X	X	.	1562	1.48	443734	248.747119	6.500000	.
3	X	X	.	X	49669	47.10	404283	202.864342	.	52.342004
4	X	X	.	.	45166	42.83	294105	135.881127	.	.
5	X	.	X	X	6	0.01	845623	.	5.666667	418.000000
6	X	.	.	X	20	0.02	640737	.	.	33.050000
7	X	.	.	.	36	0.03	545237	.	.	.

Notice in Figure 9.3 that each group from one to seven is the unique combination of missing and non-missing data values of the four variables listed. The variable EST_SPEND does not contain any missing values in the data set. For each unique group listed, the group means is listed at the right. The most important features are the frequency and percent columns in the middle of the output. The missing data pattern for these variables is not a monotone missing pattern as given in Figure 9.1. The missing data pattern can aid in your initial strategy for data analysis. If the only set of missing data is in groups five through seven, then this data consists of about 0.5% of the 105,465 records in the data set. However, a closer look at variables Prod_A and Prod_B indicate that about 43% of the data records contain no values for both product A and B combined, listed in Group 4.

9.3 Effects of Missing Data on Clustering

We will now investigate how missing data can affect and impact your clustering and segmentation analysis. We used the NYTOWNS data set back in Chapter 6, "Clustering of Many Attributes," and we will take a look at it again here. This data set did not have too many missing data elements; however, I've made an additional data set called NYTOWNS_WMISSING to have some missing data on the variables we will use to cluster the observations. However, before we do that we will need a clustering defined with the complete set of data.

Process Flow Table 2: Clustering with Missing Data

Step	Process Step Description	Brief Rationale
1	Create a new process flow diagram called Effects of Missing Data on Clustering. Add the data source NYTOWNS to diagram.	Changes PENETRATION to the target in the Input Data source node.
2	Drag a Variable Selection node and change the settings as shown in Figure 9.4.	
3	Add a Metadata node and change PENETRATION back to input.	Uses the target only for variable selection.
4	Run the flow diagram from the Metadata node, and view the Variable Selection results when complete.	Shows the analysis of variable selection with respect to PENETRATION.
5	Drag a Cluster node and attach the Metadata node to it using the settings in Figure 9.6.	Clusters NY towns from the variables selected.
6	Add NYTOWNS_WMISSING data to the process flow diagram.	Set up for identical clustering with missing data.
7	Copy the Variable Selection node and paste it on diagram.	Set up for identical clustering with missing data.
8	Copy the Metadata and Cluster nodes and paste them in the diagram.	Set up for identical clustering with missing data.
9	Change the scoring imputation in the second Cluster node to seed of nearest cluster."	Set up for identical clustering with missing data.
10	Add two more copies of Cluster nodes and change settings Omit and Mean for imputation settings.	Shows how clustering is affected by different attributes for missing values.
11	Add a SAS Code node to run the PROC MI SAS statements.	Performs multiple-imputation analysis.
12	Create a new process flow diagram called County Clusters.	Clusters NY towns geographically.
13	Drag a new Cluster node and connect NYTOWNS to it.	Clusters county, latitude, and longitude.
14	Add NYTOWNS_WMISSING data and a Score node.	Scores county clusters on missing data.
15	Add PROC MI SAS statements for imputing the missing data.	Shows multiple data imputation.
16	Understand what the SAS statements are performing.	
17	Review the cluster distance plots from the EMReport macro in the SAS code node.	Shows how distances were affected by missing values and how the MI imputed compared to the original data with no missing values.
18	Add NYTOWNS_IMPUTED data from the SAMPSIO library.	
19	Copy and paste the first Variable Selection and Cluster nodes.	

Step 1: Create a new process flow diagram and call this one Effects of Missing Data on Clustering. Drag onto it the NYTOWNS data set and change the variable called PENETRATION to a role of Target instead of Input. **Step 2:** Now, attach a Variable Selection node and connect the NYTOWNS data set to it. The Variable Selection node is useful for rapidly determining which set of variables to use when the number of potential variables is large. The NYTOWNS data set contains about 250 variables, many of which are correlated to each other. With the PENETRATION variable defined as a target variable, the Variable Selection node will attempt to set only those variables with statistical significance in explaining the target variable. It will pass all variables on to the next node; however, only the variables with high enough significance will be set to input. The settings to use for the Variable Selection node are shown in Figure 9.4. **Step 3**: Next, attach a Metadata node to the output of the Variable Selection node and edit the variables. Change the New Role of the PENETRATION variable back to Input instead of Target. **Step 4:** Run the diagram from the point of the Metadata node and when complete, open the results of the Variable Selection node. The Variable Selection window can be sorted by Role, thereby showing the results of which variables are now set to input. Also, notice that some new variables were created in the Variable Selection node. The variable AOV16_HouseMultiFamily is now an ordinal variable with several levels. Figure 9.5 shows the results of the Variable Selection node's output.

Figure 9.4 Variable Selection Node Property Settings

Property	Value
Node ID	Varsel
Imported Data	...
Variables	...
Max Class Level	100
Max Missing Percentage	50
Target Model	Default
Hide Rejected Variables	Yes
Reject Unused Variables	Yes
Chi-Square Options	
Number of Bins	50
Maximum Pass Number	6
Minimum Chi-Square	3.84
R-Square Options	
Maximum Variable Number	30
Minimum R-Square	0.0050
Stop R-Square	5.0E-4
Use AOV16 Variables	Yes
Use Group Variables	No
Use Interactions	Yes

Figure 9.5 Variable Selection Results

Name	Role	Level	Type	Label
AncItalian	Input	Interval	N	ANCESTRY (single or multiple); T...
AOV16_HouseMultiFamily	Input	Ordinal	N	AOV16: HouseMultiFamily
EduHSDip	Input	Interval	N	Educational attainment; Populatio...
IndAgric	Input	Interval	N	Employed civilian population 16 y...
JobAgriculture	Input	Interval	N	Employed civilian population 16 y...
JobOfficeSales	Input	Interval	N	Employed civilian population 16 y...
MarFemaleDivorcees	Input	Interval	N	% of marriage aged females who ...
MortageLT500	Input	Interval	N	Percent of mortgages with payme...
MortgageLT700	Input	Interval	N	Percent of mortgages with payme...
ValueLT50K	Input	Interval	N	% value less than $50,000
AncArab	Rejected	Interval	N	ANCESTRY (single or multiple); T...
AncCzech	Rejected	Interval	N	ANCESTRY (single or multiple); T...
AncDanish	Rejected	Interval	N	ANCESTRY (single or multiple); T...

Step 5: Now, attach a Cluster node to the Metadata node and change the advanced property sheet settings to those shown in Figure 9.6. The intent here is to cluster the NY towns into groups that are associated in varying degrees to the called PENETRATION variable. What we've done so far is to select the variables that appear to explain or predict the PENETRATION variable, then we switched the role for PENETRATION from a target response to an input. Now, run the Cluster node. If you open the Cluster node variables settings, you should see only the same set of variables as in Figure 9.5 set to a role of input; the remainder of the variables should have a role setting of rejected.

Figure 9.6 Cluster Node Property Sheet Settings

Property	Value
Variables	...
Cluster Variable Role	Segment
Internal Standardization	Range
Number of Clusters	
Maximum Number of Clusters	10
Specification Method	Automatic
Selection Criterion	
Clustering Method	Ward
Maximum	50
Minimum	2
CCC Cutoff	3
Encoding of Class Variables	
Ordinal Encoding	Rank
Nominal Encoding	GLM
Initial Cluster Seeds	
Seed Initialization Method	Default
Minimum Radius	0.0
Drift During Training	No
Training Options	
Training Defaults	Yes

When you run the Cluster node, the Results window should contain five clusters. The distance plot should look identical to the one in Figure 9.7.

Figure 9.7 Clustering Output Results: Distance Plot of Five Clusters

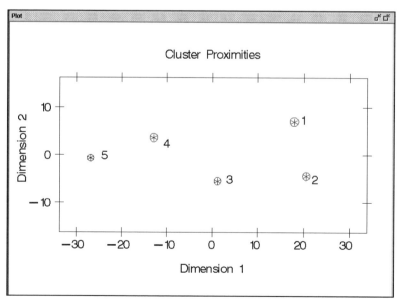

If you were to attach a Segment Profile node or just review the profiling sections in the Cluster node Results windows, then you should see that each of the five clusters contains NY towns where cluster 1 and 2 have the highest proportion of product penetration, and clusters 3 through 5 decrease accordingly. At the same time, the educational attainment variable called EDUHSDIP is strongly associated to the PENETRATION variable. Therefore, clusters with high product penetration also contain high educational attainment and vice versa. This will be the clustering analysis we will try to mimic when some of the variables we used contain missing data elements.

Now, let us see what happens when we try the same cluster settings with the same set of variables selected but with some of the values missing. **Step 6:** Underneath the process flow that you've just created, drag the NYTOWNS_WMISSING data set onto your workspace. This data set is identical to the NYTOWNS, except that the variables PENETRATION, EDUHSDIP, and ANCITALIAN now contain some level of missing data. **Step 7:** Now, copy the Variable Selection node and paste it onto the diagram. You should see a (2) after the name Variable Selection. This is a handy way to copy settings in another portion of your process flow. **Step 8:** In the same way, copy the first Metadata node and the Cluster node and paste them in the workspace; attach the data source NYTOWNS_WMISSING to the second Variable Selection node, and attach this node to the second Metadata node, and then to the second Cluster node.

After you run this process flow, you should notice that the second Variable Selection node's results are slightly different from the first node's results. This is due to the missing data elements; the Variable Selection is attempting to analyze which variables have the largest impact on the target variable Penetration. However, now with some data values missing, the analysis is altered somewhat compared to when no data was missing. If you contrast the output in Figure 9.8 with that of Figure 9.5, a few differences arise. The variable WorkClassSelf now appears in the input and EduHSDip is not listed. Also, JobConstruct replaces the JobOfficeSales. The variable FamIncLT50K appears as well in the second Variable Selection's results but not in the first one. These and other differences will cause the clustering algorithm to cluster observations differently.

Figure 9.8 Second Variable Selection Results with Missing Data

Name	Role	Level	Type	Label
AncItalian	Input	Interval	N	ANCESTRY (single or multiple); Total ancestries report...
AOV16_HouseMultiFamily	Input	Ordinal	N	AOV16: HouseMultiFamily
FamIncLT50K	Input	Interval	N	Income in 1999; Families; Less than $50,000; Percent
IndAgric	Input	Interval	N	Employed civilian population 16 years and over; Industr...
JobAgriculture	Input	Interval	N	Employed civilian population 16 years and over; Occup...
JobConstruct	Input	Interval	N	Employed civilian population 16 years and over; Occup...
MarFemaleDivorcees	Input	Interval	N	% of marriage aged females who are divorced
MortgageLT700	Input	Interval	N	Percent of mortgages with payment less than $700
ValueLT50K	Input	Interval	N	% value less than $50,000
WorkClassSelf	Input	Interval	N	Employed civilian population 16 years and over; Class ...
AncArab	Rejected	Interval	N	ANCESTRY (single or multiple); Total ancestries report...
AncCzech	Rejected	Interval	N	ANCESTRY (single or multiple); Total ancestries report...

Step 9: After you connect the second Cluster node, change the Scoring Imputation Method in the Missing Values section of the property sheet to Seed of Nearest Cluster. This change will not impute any missing data values; however, when a missing value occurs, the seed of the closest cluster will be used in its place. Notice now that the clustering results indicate only four clusters instead of the original five. The cluster distance plot of this analysis is shown in Figure 9.9.

Step 10: Now, attach two identical copies of the first Cluster node to this same process flow and change the settings in the missing values section to Omit for the nominal, interval, and ordinal types of variables. You can copy a node by highlighting the node and right-clicking to select copy. When you run this diagram, the distance plot results should be very similar to that shown Figure 9.9. If there were missing data that caused an entire record to be dropped using the fields of interest, then those records with the Omit option would drop those records entirely from the cluster analysis. Now, in the third Cluster node, set the missing section to Mean instead of Omit, and in the ordinal variable select Median. When you run this clustering, Figure 9.10 gives yet a slightly different set of clusters, even though there are still four clusters. Your process flow diagram should now look like the one in Figure 9.11.

Figure 9.9 Second Cluster Distance Plot: Missing Entries Default Settings

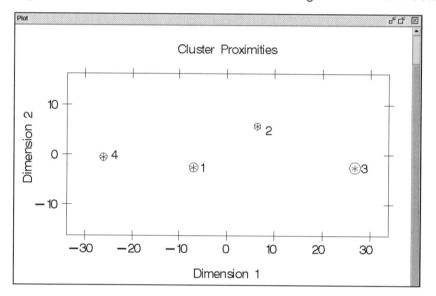

Figure 9.10 Fourth Cluster Distance Plot: Imputing Mean Values for Missing Data

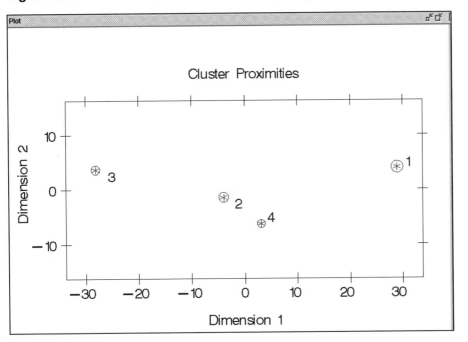

Figure 9.11 Cluster Process Flow Using Four Clustering Flows

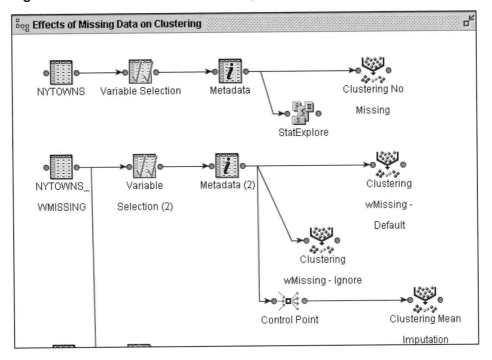

Now, we will impute some missing values on the NYTOWNS data set using PROC MI. Drag a SAS Code node to the diagram and connect the NYTOWNS_WMISSING data source to the SAS Code node. PROC MI performs multiple imputation.

9.4 Methods of Missing Data Imputation

We've looked briefly at what happens when missing values exist in clustering and have taken some efforts at circumventing the ill effects of missing values using either the seed of the nearest cluster or mean imputation in the options available inside the Cluster node. The Impute node has other data imputation options such as decision trees, with or without surrogates. We will now turn our attention to *multiple imputation* (MI).

Multiple imputation (Little and Rubin 2002) allows you to generate multiple complete data sets from data that contains missing values by replacing the missing entries with imputed ones. The imputation procedure allows an analysis on each of the imputed data sets just as if the missing entries were not missing, i.e., in the same manner as *complete case* analysis discussed earlier. The process of performing a multiple imputation involves imputing a missing entry in a data set using m multiples of imputations. This will then produce m multiple data sets where the missing entries are not complete with imputed values. One common method is to combine the m imputed data sets into a single data set that represents the all of the observed and imputed values with the added benefit of some understanding of uncertainty in the data set as a result of incorporating estimates of variability due to m number of imputations. MI actually performs simulation of the distributions of observed data used in predicting the imputed values.

When doing any kind of imputation, there are typically two competing and sometimes diametrically opposed issues to deal with. On the one hand, you want the imputed values to be *plausible* in the sense of being amply like those values that would have been there if they weren't missing since you are intending to use the imputed values as *filled-in* in your data set. Therefore, you would not want the imputed values to be out of range of the possible non-missing data elements or if you data were numeric and continuous, you would not want the imputed values to be discrete (Ake 2005). On the other hand, when using the imputed data for various analyses, you will want the results of these analyses to be unbiased. The term *unbiased* is used by statisticians when your data causes the analysis that you are attempting to perform to always under or over predict. Sometimes, meeting both of these objectives simultaneously is not always possible, and a business decision will need to be made depending on the manner in which the analysis is to be used.

If we were to impute a single imputation for each missing data element, then random variation is not present and thus bias will likely be present as I just mentioned. Without a random component, deterministic imputation methods generally will produce means and variances, which are underestimated (Little and Rubin 2002, p. 28). Random imputation can eliminate the biases that are prevalent in deterministic imputation, but some complications remain. If we use the imputed data as if it were real data, the resulting standard error estimates (variability) will be too low. The solution, at least with random imputation, is to repeat the process and is therefore why the algorithm is called *multiple imputation*. Because of the random component, the parameters governing the simulated distribution will be slightly different for each imputed data set.

In order to understand how the MI algorithm works, and the methods that you will need to employ for your missing data analysis and imputation, it is necessary to digress just a bit. In the late 1970s, Dempster, Laird, and Rubin (1997, pp. 1–38) formalized a computational method for efficient estimation of incomplete data called *expectation maximization* (EM). Around this time period, statisticians started viewing missing data in a different light than in previous years. Instead of viewing missing data as a mere nuisance, they began to see the missing values as a source of variability that could be averaged. The EM algorithm performs this averaging technique. Multiple imputation carries out the averaging via simulation; however, the SAS PROC MI, can use the EM algorithm as the initial starting points for MI (Schafer and Olsen 1998, pp. 545–571). If your missing data pattern is a monotonic one as in Figure 9.1, then certain algorithms can be applied for performing MI. However, if not, then other algorithms must be used. *Data augmentation* (DA) is another technique that uses the EM as starting points. This algorithm is called a Markov Chain Monte Carlo (MCMC) algorithm. Before we start using the DA, it is necessary to select the variables for the imputation process. You should include all possible variables that you deem sufficient in the analysis at hand, so you'll need to select variables that the business needs and ones that are highly correlated to the variables with missing data. These correlations will be valuable to the algorithm at building a predictive distribution of the variables that contain missing data. When selecting the variables and the number of iterations you need in order to use the SAS MI procedure, there are a few principles to keep in mind. First, the larger the proportion of missing data, the more iteration will be needed to reach convergence. Rubin (1987, p. 114) provides a handy formula, Equation 9.1 that estimates the relative efficiency of an estimate based on *m* imputations is given by:

$$RE = \left(1 + \frac{\lambda}{m}\right) \qquad (9.1)$$

The preceding equation indicates that if λ (the fraction of missing information) the more imputations that *m* will be required to keep the relative efficiency high. For most typical applications, three to five imputations are usually sufficient to give relative efficiencies at 80% or higher. PROC MI in SAS will compute the relative efficiency for each variable selected in the analysis.

So now, let's see if in our exercise in Process Flow Table 2 we use PROC MI using the SAS Code node to impute the values that are missing in the three variables PENETRATION, EDUHSDIP, and ANCITALIAN. **Step 11:** In the Effects of Missing Data on Clustering process flow diagram, attach a SAS Code node to the data source called NYTOWNS_WMISSING. I called this node Impute w. Proc MI. Set this diagram aside for a moment, and we'll come back to it shortly. In order to best impute the missing data, will choose to create some additional variables that will represent the geographical nature of the towns. We have counties; however, if you look at the distribution of towns within counties, it is not very evenly distributed, although each county has roughly the same amount of land area. So, we could cluster the towns by their geographical location as we do have their latitude and longitude on this data set. **Step 12:** Create a new process flow diagram; I called it County Clusters. **Step 13:** Open the NYTOWNS data set (this one does not have the missing values) and attach a Cluster node to it. The only variables you should select are COUNTY, LATITUDE, and LONGITUDE. All other settings in the Cluster node property sheet should be default settings. This should generate six county clusters and keep the numbers of towns reasonably consistent. **Step 14:** Now attach another copy of the NYTOWNS_WMISSING data source and a Score node so that your diagram looks like the one in Figure 9.12. The Score node will generate scoring results of the clusters generated so we can now use the newly generated cluster segments.

Figure 9.12 County Clusters Process Flow Diagram

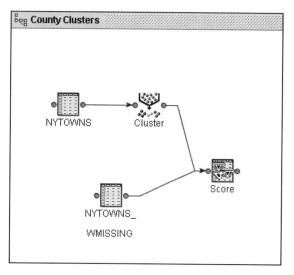

In the Score node, set the type of scored data property to Data instead of View. This will generate a SAS data set with the scored cluster segments instead of creating a SAS data view. Run the process flow from the Score node. Your county cluster distance plot should look like the one in Figure 9.13.

Figure 9.13 Cluster Distance Plot of County Clusters

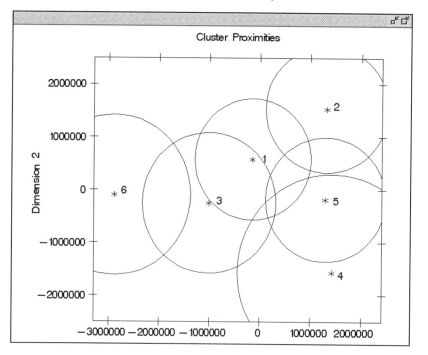

Now, using our original process flow diagram, which is shown in Figure 9.14, you can open the SAS Code node and begin to use PROC MI to perform multiple imputation. We will also use some of the SAS Enterprise Miner automated macros, which will allow us to create plots and allow them to be displayed as menu options within the results menus in the SAS Code node Results window. Your process flow diagram should look like the one in Figure 9.14.

Figure 9.14 Current Process Flow Diagram with SAS Code to Run the PROC MI Procedure

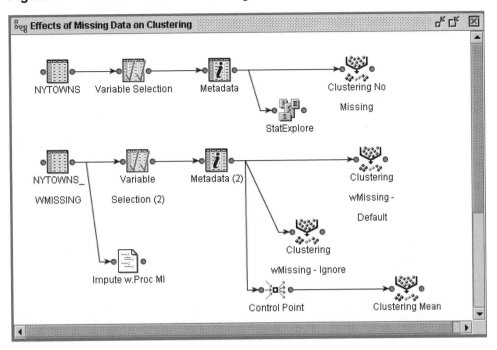

Step 15: Open the SAS Code node (which I labeled Impute w. Proc MI) and insert the SAS code as given in Figure 9.15. The SAS statements are in Appendix 3 in the Chapter 9 folder of your data CD. The file should be named PROC_MI.SAS. Now, let's go over this code to review what is happening at each stage so you can assimilate it and begin to use and appreciate what it is being performed.

Figure 9.15 SAS Code for the PROC MI

```
/* Take scored cluster results from County Clusters diagram and sort
by cluster*/
/* segment */
proc sort data=sampsio.nytowns_county;
 by _segment_ ;
 run;
ods html
body='c:\temp\proc_mi.htm' style=BarrettsBlue;
title 'Missing Imputation Analysis';
 proc mi data=sampsio.nytowns_county ❶
    seed=1234567 nimpute=5 minmaxiter=200 min=. . . . 0    6    0
        ❷              ❸               ❹           max=. . . . 48.5 59.5 44.8
    out=work.total_impute ; ❺
❻mcmc chain=single displayinit initial=em(itprint) impute=full
    outest=work.mcmc_est ;
❼  by _segment_ ;
❽var mortgagelt700 valuelt50k indagric marfemaledivorcees
  penetration eduhsdip ancitalian;
  run;
  ods html close;
❾proc summary data=work.total_impute mean nway ;
   class geo_id;
  var penetration eduhsdip ancitalian ;
  output out=work.impute_means mean= ;
  where missing_flag=1;
  run;
```

```
      proc sort data=work.impute_means ;
        by geo_id; run; ❿
      proc sort data=sampsio.nytowns_county out=work.orig_ds;
      by geo_id;
      run;
      data sampsio.nytowns_imputed;
        merge work.orig_ds(in=a)
        work.impute_means(in=b drop=_freq_);
      by geo_id;
        if a;
      run;

❶❶proc means data=sampsio.nytowns mean stderr noprint;
        var penetration eduhsdip ancitalian ;
      output out=work.orig_mean mean= stderr=stderr(penetration)=stde_penet
            stderr=stderr(eduhsdip)=stde_eduh
            stderr=stderr(ancitalian)=stde_anci ;
      run;
❶❶proc means data=sampsio.nytowns_imputed mean stderr noprint;
        var penetration eduhsdip ancitalian ;
      output out=work.impute_mean mean=
      stderr=stderr(penetration)=stde_penet
          stderr=stderr(eduhsdip)=stde_eduh
            stderr=stderr(ancitalian)=stde_anci ;
      run;

      /* Now format the output of the Mean-StdErr data set */
      ods html
      body='c:\temp\mi_compare.htm' style=BarrettsBlue;
      title 'Imputed Means & Std.Errors on Imputed Data set';
❶❷proc sql;
      select _freq_ label='# of Obs',
          penetration label='Mean Penetration',
          eduhsdip label='Mean Edu Attainment',
          ancitalian label='Mean Italian Ancestry',
          stde_penet label='SE Penetration',
          stde_eduh label='SE Edu Attainment',
          stde_anci label='SE Italian Ancestry'
          from work.impute_mean
        ;
      quit;
      title 'NYTowns Means & Std.Errors on Original Data Set';
❶❷proc sql;
        select _freq_ label='# of Obs',
          penetration label='Mean Penetration',
          eduhsdip label='Mean Edu Attainment',
          ancitalian label='Mean Italian Ancestry',
          stde_penet label='SE Penetration',
          stde_eduh label='SE Edu Attainment',
          stde_anci label='SE Italian Ancestry'
          from work.orig_mean
        ;
      quit;
      ods html close;

      %em_register(key=first,type=data); ❶❸
      %em_register(key=second,type=data);
      %em_register(key=third,type=data);
      data &em_user_first;
        set emws.clus_train;
      run;
```

```
data &em_user_second;
  set emws.clus4_train; ⓮
run;
data &em_user_third;
    set emws.clus5_train; run; ⓮
proc sort data=first; by _segment_ ;
proc sort data=second; by _segment_;
proc sort data=third; by _segment_;
run;
⓯%em_report(key=first,viewtype=histogram,x=distance,block=Plots, ⓰
  by=_segment_,description=Orig Clustering Distances);
  %em_report(key=second,viewtype=histogram,x=distance,block=Plots, ⓰
  by=_segment_,description=Mean Imputation Cluster Distances);
  %em_report(key=third,viewtype=histogram,x=distance,block=Plots, ⓰
  by=_segment_,description=MI Imputed Cluster Distances);
```

Step 16: Here is an explanation of what the SAS statements perform:

❶ PROC MI statement includes a Data=, which points to the data set that contains the missing values we would like to impute.

❷ The SEED= is the random seed option so that if you were to run the program on a different date or time, you would still get the same results as any random selections are starting with the same initial seed value.

❸ NIMPUTE=5 tells the PROC MI that we want to have five completed data sets of imputations runs. That is, we want five complete data sets replicated with all variables and rows, except that the missing values of each of the five runs are different random draws for the simulated distribution of each variable to be imputed.

❹ MINMAXITER=200 means the maximum number of iterations will be 200 within the specified range, which is specified by the Minimum and Maximum parameter settings. The minimum and maximum settings for the three variables that have missing values (PENETRATION, EDUHSDIP, and ANCITALIAN) are 0, 6, 0 for the minimum values and 48.5, 59.5, 44.8 for the maximum values. The preceding decimal points indicate that the variables listed in the VAR statement don't have a minimum or a maximum range. These variables don't even have any missing entries; however, they are being used to help predict the missing values for the variables PENETRATION, EDUHSDIP, and ANCITALIAN.

❺ The OUT= option specifies the output data set that will contain all the imputed runs. A new variable on that data set called _IMPUTATION_ will be the imputation number for that data set.

❻ The MCMC statement indicates that the Markov Chain Monte Carlo method will be used to perform the random draws for the imputations. This algorithm is a simulation of random variables that creates a Markov chain. The chain is a sequence of random variables in which the distribution of each element depends only on the value of the previous one. In MCMC simulation, one constructs a Markov chain long enough for the distribution of the elements to stabilize to a stationary distribution, which is the distribution of interest. By repeatedly simulating steps of the chain, the method simulates draws from the distribution of interest. In the MCMC statement, I selected a single Markov chain; however, multiple ones can be selected as well.

❼ The BY statement (by _segment_) means that a separate imputation analysis will be performed for each segment.

❽ The VAR statement then lists the variables to be imputed. Pay particular attention to the ordering so that you specify the minimum and maximum ranges for each variable in the same sequence as in the VAR statement.

Well, so much for the PROC MI statements. What follows is a summarization step with PROC SUMMARY ❾ that computes the mean of each imputation in the data set created from PROC MI where the missing flag equals 1. Then, the original data set and the imputed averages are merged into a single data set by each record ID (geo_id) ❿. This effectively combines the average imputed values for the five imputed runs into the original data set that contained the missing values; this produces a complete data set where the original data and the imputed values from missing data are merged. To compare the data set that had missing values with the original NYTOWNS data set, two sets of PROC MEANS are run ⓫ to estimate the mean and standard errors for the three variables of interest (PENETRATION, EDUHSDIP, and ANCITALIAN). The PROC SQL statements ⓬ format the selected variables from the PROC MEANS output.

The following statements are macros that are provided in SAS Enterprise Miner to register data sets ⓭. Then, the cluster data sets ⓮ are written and sorted in the SAS Enterprise Miner data library for the project using automated macros provided in the SAS Code node. The next set of statements use the %EM_Report macro ⓯. This macro is a handy tool and one of the SAS Code node utility macros provided. The %EM_Report macro allows you to register plots within the SAS Code node results area. The plots can then appear as menu items. I added three histogram plots ⓰: one for the cluster distances in the original NYTOWNS clustering, one for the distances in the mean imputation cluster analysis, and one in the imputed with PROC MI cluster analysis. **Step 17:** Figure 9.16 shows the menu with the plots.

Figure 9.16 SAS Code Node Results with Custom Plots

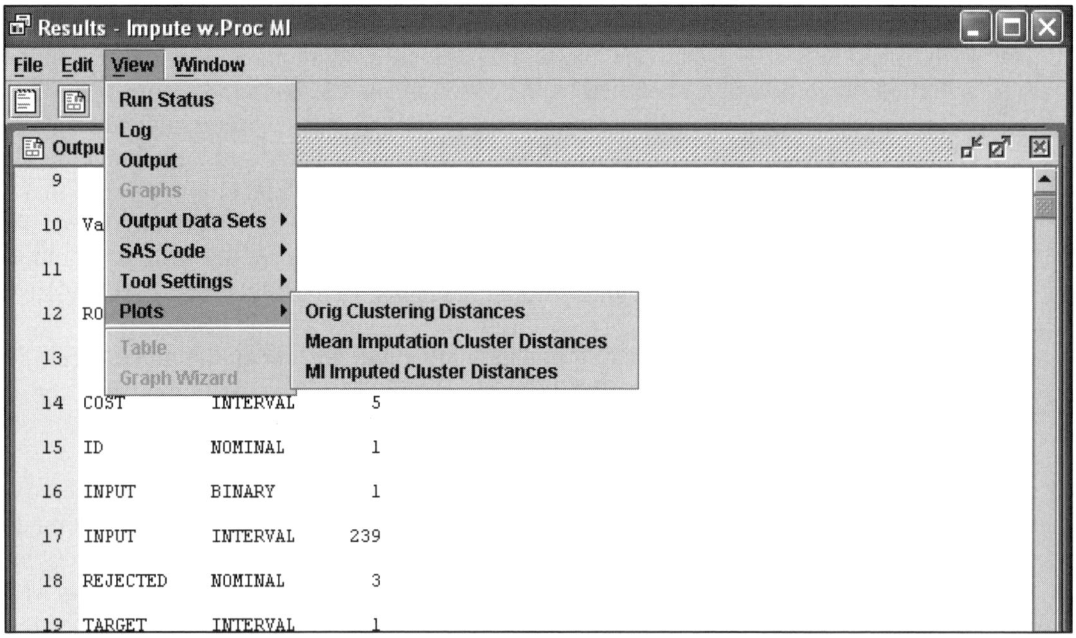

Each of the three plot selections in Figure 9.16 shows the distances obtained from clustering that are to be graphed as histograms grouped by each cluster segment. Comparison of the original cluster distances to the data sets where mean and multiple imputations will allow us to see how well the imputations performed on the missing values, as the cluster settings in the Cluster node were identical. So, lets look at these histograms to see if we can infer anything from our analysis. Figure 9.17 shows the comparison of the original cluster distances with the mean imputation cluster distances, and Figure 9.18 shows the comparison of the original cluster distances with the multiple imputation cluster distances.

Figure 9.17 Comparison Histograms of Original vs. Mean Imputation Cluster Distances

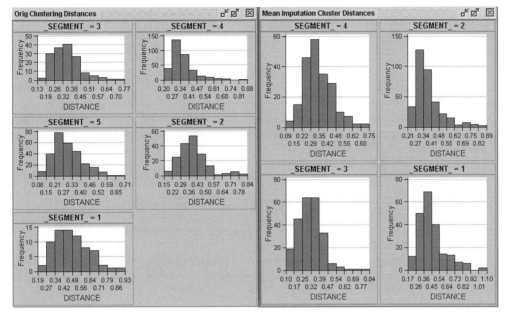

Figure 9.18 Comparison Histograms of Original vs. Multiple Imputation Cluster Distances

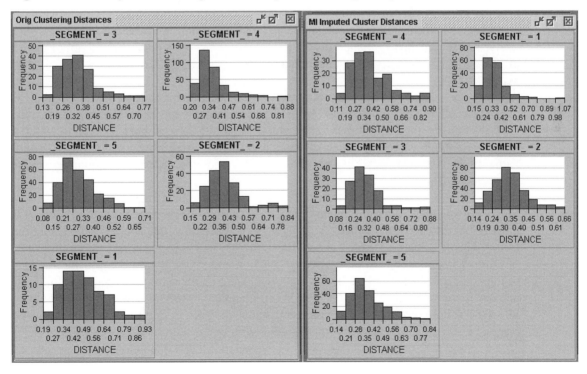

The clustering results in the mean imputation data run produced only four clusters, as shown in Figure 9.10. The distances are rather different from the original. The multiple imputation distances are in general much more in line with the original. The output tables from PROC MEANS show the mean and standard errors in Figure 9.19.

Figure 9.19 Comparison Tables of Original Clusters and Multiply-Imputed Clusters (Mean and Standard Errors)

NYTowns Means & Std.Errors on Original Data Set

# of Obs	Mean Penetration	Mean Edu Attainment	Mean Italian Ancestry	SE Penetration	SE Edu Attainment	SE Italian Ancestry
1006	8.6005964215	36.210934394	11.297614314	0.2284727465	0.2334424463	0.2140998203

Multiply   Imputed Means & Std.Errors on Imputed Data Set

# of Obs	Mean Penetration	Mean Edu Attainment	Mean Italian Ancestry	SE Penetration	SE Edu Attainment	SE Italian Ancestry
1006	8.7532851883	36.337524228	11.330032414	0.2285900593	0.2368605361	0.2138557959

Notice that the mean and standard errors on the multiply-imputed data are rather close to the original data set. This shows that the overall distributions did not change too much and that little or no bias is in the multiply-imputed data set, which is what we strove for in the first place. Any analysis that uses the aggregated results to infer some business decision will not be adversely affected by biased results using this technique.

Although the mean imputation is not quite as good, the Impute node in SAS Enterprise Miner does allow other types of means to impute, such as Tukey's biweight, Huber's, and Andrew's wave, which are good robust mean estimators. What they lack is that the mean in which they impute is a single value and is thus constant and has no variability associated with the imputed value. In the next section, we'll look at a method for computing estimated confidence intervals for imputed data values.

Step 18: Now that we have imputed the values, let's cluster just like in the first process flow on the original data set. On the process flow diagram, add a new data source to the Data Sources folder. The data source is from the SAMPSIO library and is called NYTOWNS_IMPUTED data set. **Step 19:** Add a copy of the Variable Selection node (now the third copy) and the remainder of the nodes just like the other flows. Be sure to keep all the settings the same as in the first process flow. Now run the Cluster node. Your completed process flow diagram should now look like the one in Figure 9.20 and the cluster distance diagram in Figure 9.21. Now compare the original cluster distance diagram in Figure 9.7 with that in Figure 9.21.

Figure 9.20 Completed Process Flow Diagram with All Flows

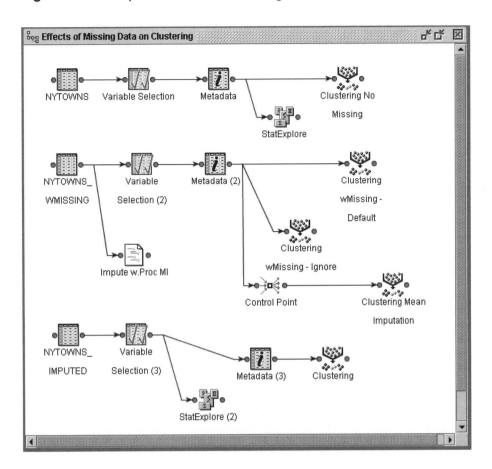

9.5 Obtaining Confidence Interval Estimates on Imputed Values

There are a variety of methods for computing confidence intervals using a technique called bootstrapping and the jackknife. *Bootstrapping* is an algorithm in which the estimates are based on dropping a single or multiple observations from the sample. The process also involves repeated resampling of the observed data (Little and Rubin 2002, pp. 79–82). Sampling in this fashion allows estimates of variability and therefore computations of the confidence intervals. To accomplish the estimates of confidence intervals, we can employ a SAS macro called jackboot.sas obtained from the SAS Web site. This macro performs jackknife and bootstrap methods to obtain one-sided or two-sided confidence intervals. The SAS Web site provides some preliminary documentation on its use and a little on the techniques of bootstrapping as well. This macro is included on the CD for this book.

A *confidence interval* is a statistic that affords us the ability to understand how much variability exists in our data. The confidence interval is an estimate that depicts a range that a point estimate must lie between at a certain level of *confidence*. For example, if we have a confidence interval of 12.5 and 18.5 at a 90% confidence and our estimate is 15.5, then we can say that we are 90%

confident that our estimate should lie between 12.5 and 18.5. Alternatively, if we performed this estimate 100 times, roughly 90 times out of the hundred the estimate should fall between 12.5 and 18.5 on average. Obviously, the more variation that exists in the data, the wider the confidence interval will become and the less confident we will be about that estimate. PROC MI provides confidence intervals on the *parameter estimates* used in the simulation of the distributions of the variables in the analysis; however, that is not the same thing as a confidence interval estimate of the imputed value.

To see how we can apply this macro, open the Program Editor window in the SAS Code node. At the bottom of the code that has been entered so far, place the statements as shown in Figure 9.21. The jackboot.sas macro is included on the CD in the Chapter 9 folder.

Figure 9.21 Additional SAS Code to Run the Bootstrap Analysis for Confidence Intervals

```
%inc 'c:\temp\jackboot.sas';
%macro analyze(data=sampsio.nytowns_imputed,out=impute_averages);
 proc summary data=&data  mean nway;
    var penetration eduhsdip ancitalian ;
output out=&out  mean= ;
 %bystmt;
run;
%mend;

%boot(data=work.impute_means,alpha=0.1,samples=100,random=123,size=10
0,
stat=penetration eduhsdip ancitalian);
```

You should either copy it to a simple location like `c:\temp` or modify the first line of the SAS statement in Figure 9.21 to place the INCLUDE statement to reference some other area where you would like to store the bootstrap macro. The additional statement running the analyze macro runs the analysis. You could modify this further (for extra credit) by replacing the %BYSTMT statement with a variable such as the SEGMENT variable that was used in the clustering of the NY towns into super counties. You would also need to alter the jackboot.sas macro at the end to reflect the %BYSTMT statement in a similar fashion in order to perform this.

The output that the jackboot macro produces is now placed in the Output window of the results in the SAS Code node. Figure 9.22 shows a portion of this output giving the confidence interval of the mean value for each of the requested variables that multiple imputations were performed.

Figure 9.22 Bootstrap Macro Output Results

	Observed	Bootstrap	Approximate	Approximate Standard	Approximate Lower Confidence	Bias-Corrected	Approximate Upper Confidence
Name	Statistic	Mean	Bias	Error	Limit	Statistic	Limit
AncItalian	11.373294175	11.417883099	0.04459	0.5136630693	10.4838	11.3287	12.1736
EduHSDip	36.514900761	36.68494499	0.17004	0.6895351413	35.2107	36.3449	37.4790
Penetration	9.05321699	9.0793807975	0.02616	0.6334947594	7.9850	9.0271	10.0691

(2246 NYTowns Means & Std.Errors on Original Data Set)

Name	Confidence Level (%)	Method for Confidence Interval	Minimum Resampled Estimate	Maximum Resampled Estimate	Number of Resamples
AncItalian	90	Bootstrap Normal	10.136168596	12.658304345	100
EduHSDip	90	Bootstrap Normal	35.213773474	38.257405238	100
Penetration	90	Bootstrap Normal	7.4391707142	10.645503288	100

I hope that in this chapter you've seen some of the effects of missing data on clustering and that this concept is extended to other data mining analyses as well. Multiple imputation is a technique that can be applied to aid in your analysis so that more data records can be used to complete the analysis.

9.6 References

Ake, Christopher F. 2005. "Rounding After Multiple Imputation With Non-binary Categorical Covariates." *Proceedings of the Thirtieth Annual SAS Users Group International Conference.* Cary, NC: SAS Institute Inc. Paper no. 112–30.

Allison, Paul David. 2002. *Missing Data.* Thousand Oaks, CA: Sage Publications, Inc.

Dempster, A. P., N. M Laird, and D. B Rubin. 1977. "Maximum Likelihood from Incomplete Data via the EM Algorihtm." *Journal of the Royal Statistical Society.* B 39.1:1–38.

Little, Roderick J. A., and Donald B. Rubin. 2002. *Statistical Analysis with Missing Data.* 2d ed. Hoboken, NJ: John Wiley & Sons, Inc.

Rubin, Donald B. 1987. *Multiple Imputation for Nonresponse in Surveys.* New York: John Wiley & Sons, Inc.

SAS Enterprise Miner, Release 5.2. Cary, NC: SAS Institute Inc.

Schafer, Joseph L., and Maren K., Olsen. 1998. "Multiple Imputation for Multivariate Missing-Data Problems: A Data Analyst's Perspective." *Multivariate Behavioral Research.* 33.4:545–571.

Product Affinity and Clustering of Product Affinities

10.1 Motivation of Estimating Product Affinity by Segment

We now come to the point in our course of study where we would like to add some new dimensions to our customer segmentation; that is, we want to add the customer's product affinities from purchases or their product interests. There are some very good reasons for adding a product dimension to our customer segmentation. In many business situations, a marketer, sales personnel, or other professional that endeavors to cross-sell, up-sell, or sell into a prospect account will need to know what the customer has already purchased or their product interests in order to increase the product portfolio of the customer and at the same time to increase their valuation. This will also increase their purchase loyalty, especially if a product or service purchase is displacing one or more of a competitor's products or services.

Understanding your customer's business will be of great value to marketers and sales professionals in your organization by knowing how your products and services help your customer's success. Some of this understanding may come in the form of knowing which items your customer has purchased together, at what times, and how often. One method for understanding this is through *market basket* analysis. *Market basket* analysis indicates that your customer has product A and product B purchased together in a specific time frame. You can think of market basket analysis as a shopping cart in a retail or grocery store with groups of items to be purchased at the checkout line. *Market basket* analysis does not refer to any one particular analytical technique or algorithm; instead, it refers to a set of business issues related to point-of-sale data transactions. One of the more common techniques in understanding *Market baskets* is to analyze these transactions with association rules (Berry and Linoff 2004, pp. 289–301). In order to generate an association rule, you need three specific data elements. First, you need a customer (or customer identification). Second, you need an order or transaction that contains items purchased. In journal literature, these purchased items are typically called *item sets*. Lastly, you need to record the actual product that was purchased. Typically, the order invoice data looks like line item records as shown below.

Invoice Date	Part Number	Customer ID	Gross Revenue	Net Revenue	Quantity
10-Oct-2001	138478-B21QS	1248493847	$165.38	$136.25	1
11-Oct-2001	384378-AK37	4738278329	$2110.44	$1935.65	1
01-Nov-2001	135537-D3841	8374382733	$300.00	$270.00	3

Pivoting this into a transaction view produces a transaction data set and when transposed again it can be merged with the customer view of data (one row per customer) to give a data set with product information and customer information in one.

Association discovery is the identification of items that occur together in a given event or record. In purchase association, items that are purchased together are associated with one another. The rules from association analysis are often expressed as follows: if item A is purchased with item B, then item C is also purchased. This is a two-level association rule. There can be one, two, three, or more rules that can be formed. The rules should not be thought of as a direct causation, but just an association between two or more items. Examples of some hypothetical association discovery rules include the following:

- If customers buys soap, then 10% of the time they also purchase paper towels.
- A grocery store may find that 80% of all shoppers will buy bean dip when they also purchase a bag of chips.
- If people visit Web page A, then they also click on Web page B.
- Investors holding an equity index fund will also have a growth fund in their portfolio 40% of the time (SAS Institute 2003a).

Confidence, level of support, and lift are three evaluation criteria of association rule discovery. The strength of the association is measured by the confidence factor. This is the percentage of cases in which a resulting event occurs given that a predecessor event also occurs. Support is how frequently the combination occurs in the market basket (database data set). Lift is the confidence factor divided by the expected confidence. Lift is the factor by which the likelihood of the resulting event occurs given the predecessor event. Figure 10.1 demonstrates the relationship among support, confidence, and lift. Association rules with high support and confidence are worthy to note; however, rules with high confidence but low support should be interpreted with extreme caution.

Figure 10.1 Venn Diagram of Product Association Metrics

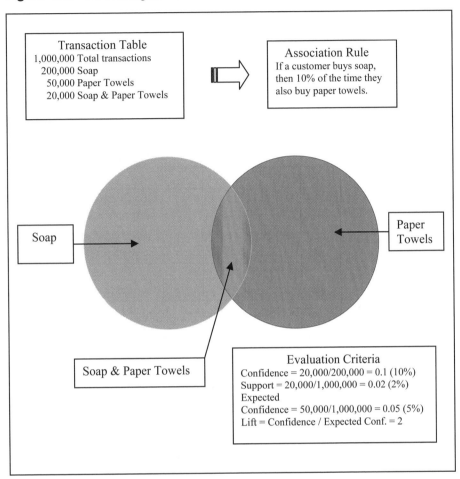

Another method is to cluster customers into similar groups that have common product sets or interests. In order to accomplish this, however, the quantities of product must be transformed, as a typical distribution of product quantities is usually rather skewed in nature. Of course, you could transform the quantities into a different distribution, but this would not allow you to compare one customer against another as each customer has different quantities of different products. The kind of transformation that is needed would be so that each customer can be compared on somewhat equal footings. For instance, let's take a hypothetical example. Customer A has purchased a quantity of 25 items all at once with one product, while customer B has purchased 25 items but in three product groups and spread over more time. Are these customers similar in their product portfolio? The answer depends on how you would measure similarity. If it is only quantity, then yes they are similar as they have the same purchase quantities; however, if you were to combine depth of product portfolio and the time element, then perhaps they are not that similar. This chapter should help you to understand some of the methods available to you for estimating product affinities by customer segment, that is, overlaying a product affinity score onto existing customer segmentation and even cluster customers according to their product affinities. We will look at some of the advantages and disadvantages of these techniques.

10.2 Estimating Product Affinity Using Purchase Quantities

Process Flow Table: Binary Product Affinity

Step	Process Step Description	Brief Rationale
1	Create a project called Product Affinities and a new process flow diagram called Binary Product Affinity.	
2	Add the CUSTOMERS data set and a SAS Code node.	
3	Explore the product quantity distributions.	Shows the data assay of the product quantity data.
4	Add the code statement into the SAS Code node.	
5	Open the Chapter 5 segmentation project; attach the Score node to the Cluster node.	Scores clustering and collects the code.
6	Add a SAS Code node to the Score node to output a scored data set.	Saves the data set to the SAMPSIO library.
7	Drag the CUSTOMER_SCORE data to the diagram and Merge node.	Shows the newly scored data with the cluster model.
8	Re-open the SAS Code node and add statements to compute binary affinities.	Shows the binary product score are means 0–1.
9	Add a Metadata node and change the binary product score's role to predict.	Changes the binary score to the predict role.
10	Add a second SAS Code node and label it Affinity by Segment.	Computes the mean binary score by segment.
11	Add a Cluster node and connect the Metadata node to it.	Shows how clustering deals with abnormal product quantity data.
12	Copy the Cluster node and paste it just below the other one.	Clusters product quantities for A, B, and C.
13	Drag a Transform Variables node and connect it as shown in Figure 10.11.	Transforms A through C and then re-clusters.
14	Set the Transform Variable node and Cluster node properties.	Shows the settings for the transformed quantities.
15	Force the Cluster node maximum number of clusters to 12 and re-cluster.	Seeing if max clusters has any affect.
16	Add a SAS Code node and connect the Metadata node to it.	Uses the softmax macro to scale prod A–C.
17	Add additional SAS statements to the SAS Code node with softmax.	Compares SAS softmax to macro.
18	Add the newly scored data from the softmax scaling SOFT_SCORE to the diagram.	Adds SOFT_SCORE from the SAMPSIO library to the Data Sources folder and diagram.
19	Copy two more Cluster nodes, one for softmax scores and the other for SAS softmax scores.	Compares SAS softmax to the macro softmax computations.

When a distribution of quantity data contains very long tails, then one possible alternative is to turn the product quantity into a different representation. If you have one or more items of product A, then represent the quantity of product A by 1; otherwise, represent it with a zero. This is advantageous when you take a mean of a column that contains only binary data (0 and 1), then the mean will also be between 0 and 1. This does, however, skirt the issue of a customer who has purchased a large quantity versus a customer who has purchased a much smaller quantity, so you could also devise a weighting scheme to go along with the binary representation. We will see the advantages of a binary representation and possibly some of the disadvantages as well. The CUSTOMER data set in the SAMPSIO library contains some product purchases labeled Prod_A through Prod_Q with several product options as well. The distribution of Prod_C quantities is shown in Figure 10.2. Notice the very long tails of product quantities and rather small percentages of customers who have purchased large quantities of items. Data of this kind will cause problems in the analysis as it current stands, especially if placed directly into a K-means clustering algorithm without any sort of transformation.

Figure 10.2 Distribution of Product C

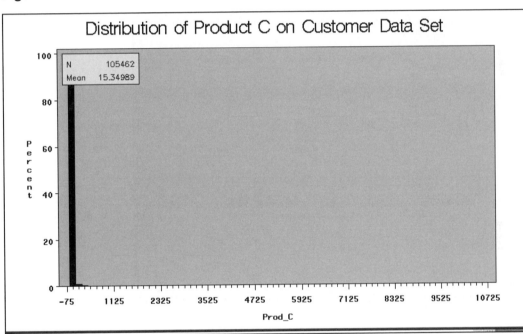

The technique of binary representation as an indicator of product purchase or non-purchase only allows one to review if the customer actually has the product or not. It also allows an assessment of product affinity in a group of customers. Let us now see how to compute these binary representations. **Step 1:** If you have not already done so, open SAS Enterprise Miner and create a new project called Product Affinities. Now, add a data source to your Data Sources folder and select in the CUSTOMERS data set located in the SAMPSIO SAS library. Be sure that all variables are selected as Input with the exception of customer ID, which should be set to a role of ID. Create a new diagram and call it Binary Product Affinity. **Step 2:** Add the newly created data source CUSTOMERS to your diagram and also add a SAS Code node to the diagram. Connect the CUSTOMERS data source to the SAS Code node. At this point, since we haven't looked at the product quantities too much, you should view the data set just to get a feel for the product data. **Step 3:** If you highlight the CUSTOMERS data source in the Data Sources folder, you can select Explore the data set by right-clicking on the CUSTOMERS icon. View the data set now and especially take note of the Prod_A through Prod_Q columns. Figure 10.3 shows a partial view of the CUSTOMERS data set. Notice that some of the rows in the product columns contain empty values. When this data set was built, the product invoice data was in a transaction format and therefore transposed and merged into a customer view (one row per customer) data set. The empty values are actually zero purchases so they should be filled with zeros. If we are going to compute an average or a sum from any of these columns, then zero is a valid number, whereas an empty value will be treated as null and therefore thrown out of the computations altogether. If you look closely enough or if you run sum basic statistics, you should also notice that the minimum is not zero but usually a negative value. These negative values are actually product quantities that were shipped to the customer and returned. We may want to treat these as a special case or perhaps just leave them as is in our computations. More about this a little later on.

Step 4: Highlight the SAS Code node and open the property sheet item SAS Code. We will first set the empty values in the product categories to zero and leave the negative values alone for the moment. Place in the Code tab the following lines of SAS Code shown in Figure 10.4. This code is located on your data CD in the Chapter 10 folder as shown in Appendix 3. This will now set and fill the product category variables to zero when empty.

Figure 10.3 CUSTOMERS Data Set View Table

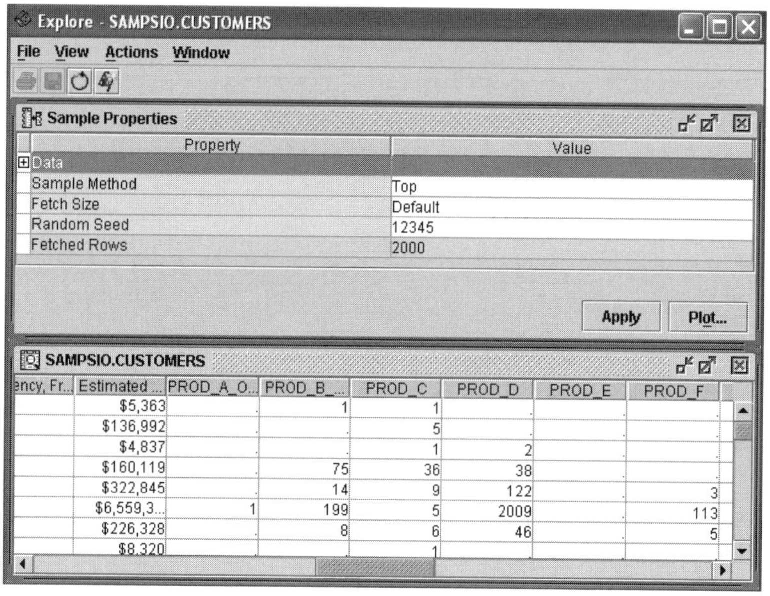

Figure 10.4 SAS Code for Binary Product Affinity: Zero Setting

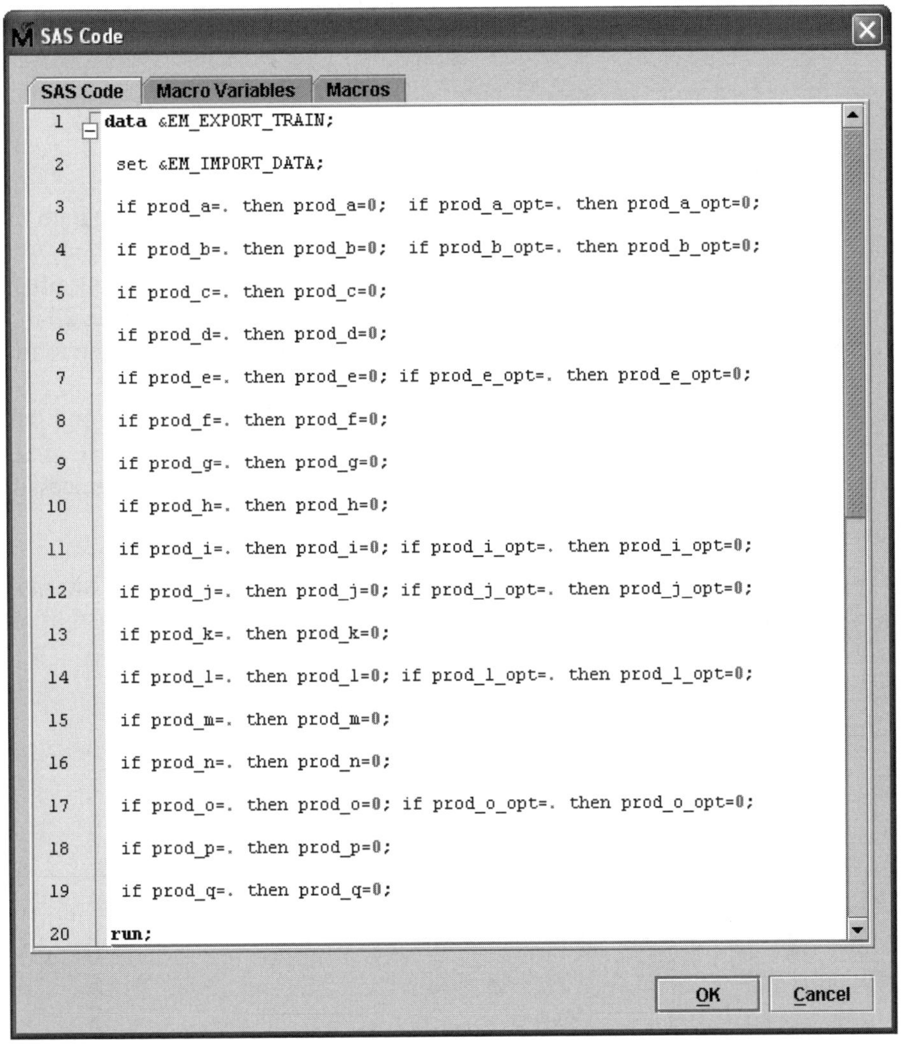

```
M SAS Code                                                    X

 SAS Code   Macro Variables   Macros

 1   data &EM_EXPORT_TRAIN;
 2     set &EM_IMPORT_DATA;
 3     if prod_a=. then prod_a=0;  if prod_a_opt=. then prod_a_opt=0;
 4     if prod_b=. then prod_b=0;  if prod_b_opt=. then prod_b_opt=0;
 5     if prod_c=. then prod_c=0;
 6     if prod_d=. then prod_d=0;
 7     if prod_e=. then prod_e=0; if prod_e_opt=. then prod_e_opt=0;
 8     if prod_f=. then prod_f=0;
 9     if prod_g=. then prod_g=0;
10     if prod_h=. then prod_h=0;
11     if prod_i=. then prod_i=0; if prod_i_opt=. then prod_i_opt=0;
12     if prod_j=. then prod_j=0; if prod_j_opt=. then prod_j_opt=0;
13     if prod_k=. then prod_k=0;
14     if prod_l=. then prod_l=0; if prod_l_opt=. then prod_l_opt=0;
15     if prod_m=. then prod_m=0;
16     if prod_n=. then prod_n=0;
17     if prod_o=. then prod_o=0; if prod_o_opt=. then prod_o_opt=0;
18     if prod_p=. then prod_p=0;
19     if prod_q=. then prod_q=0;
20   run;
                                              OK      Cancel
```

Notice that the data set names are SAS macro variables that are created by default in the SAS Enterprise Miner SAS Code node. This is helpful in developing data-driven data mining applications without hard coding the actual data set names (not that that is wrong) and allows the application to be more transferable without a lot of rewriting.

10.3 Combining Product Affinities by Cluster Segments

Step 5: Remember the B-B Segmentation diagram, in Chapter 5, "Segmentation of Several Attributes with Clustering?" We used the CUSTOMERS data set to perform a B-B segmentation and segment profile. With this SAS Enterprise Miner session still in use, double-click the EM Client icon on your desktop. This will start another client session and you can open another project at the same time. You just can't re-open the same project and diagram you already have open. Open the Segmentation Example project and the B-B Segmentation diagram. Place in the clustering process flow a Score node to capture all the clustering and data pre-processing so we can place the segment clusters onto our product affinity diagram process flow. Place a Score node in your B-B Segmentation process flow diagram (Figure 5.7) and connect the output of the Cluster node to the Score node. Also, drag the CUSTOMERS data source onto the project flow and change the role of this data to Score. **Step 6**: Also, attach a SAS Code node after the Score node and place the following code in it to save the data set into the SAMPSIO SAS library. This will copy the scoring data set into the SAMPSIO library for us to use in this project.

```
DATA SAMPSIO.CUSTOMER_SCORE;
SET EMWS.SCORE_SCORE;
RUN;
```

Now, run the Score node. To view the SAS code, use the View menu by selecting Path Publish Score Code. In the Data Source folder, add a data source and copy the data set called CUSTOMER_SCORE in the SAMPSIO library. Drag this scored data set onto your process flow diagram and in the Input Data node properties, open the variables property, and set all the product variables to Drop=Yes. This will allow only the product variables in our Code node to filter through. **Step 7**: Connect the CUSTOMER_SCORE node and the SAS Code node to a Merge node. Set the Merge node properties so that type of merging is MATCH MERGE and in the variables property, set the CUST_ID to the BY variable in which to merge both data sets. Set all other variables to DROP except for _SEGMENT_ variable and CUST_ID. Your process flow diagram should now look like the one in Figure 10.5.

Figure 10.5 Product Affinities Process Flow with Segment Scoring and Data Merging

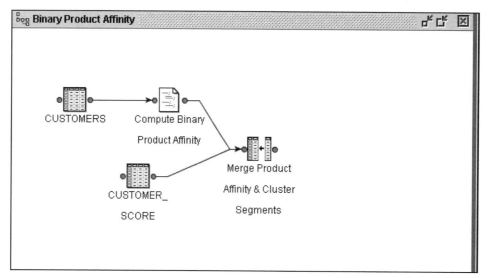

The settings in the Merge node should indicate that one-to-one match merging is selected. Open the Variables property and set the MERGE ROLE on the CUST_ID to BY. This will match merge the product data with zero fills for null values and the scored customer segmentation data for each customer ID. The data sets are both internally sorted on each BY variable. Now run the Merge node. When you view the resulting data set from the Merge node, your data should now contain the cluster segment variables _SEGMENT_ and DISTANCE and also the product values we filled in with zeros for empty values.

The idea now is to compute estimates of average product quantity for each product on each segment. We will also compute the binary product affinities and contrast and compare the results of these two methods. So, to compute the binary product scores from raw quantities, we need to translate 0 for 0 quantity and 1 for any quantity greater than or equal to 1. **Step 8:** Reopen the SAS Code node and at the bottom just before the run statement, and place the following set of SAS code to create binary product scores as shown in Figure 10.6.

Figure 10.6 Additional SAS Code to Compute Binary Product Affinities

```
/* Compute binary product scores from quantities */
   if prod_a=0 then bin_a=0; else bin_a=1;
   if prod_a_opt=0 then bin_a_opt=0; else bin_a_opt=1;
   if prod_b=0 then bin_b=0; else bin_b=1;
   if prod_c=0 then bin_c=0; else bin_c=1;
   if prod_d=0 then bin_d=0; else bin_d=1;
   if prod_e=0 then bin_e=0; else bin_e=1;
   if prod_e_opt=0 then bin_e_opt=0; else bin_e_opt=1;
   if prod_f=0 then bin_f=0; else bin_f=1;
   if prod_g=0 then bin_g=0; else bin_g=1;
   if prod_h=0 then bin_h=0; else bin_h=1;
   if prod_i=0 then bin_i=0; else bin_i=1;
   if prod_i_opt=0 then bin_i_opt=0; else bin_i_opt=1;
   if prod_j=0 then bin_j=0; else bin_j=1;
   if prod_j_opt=0 then bin_j_opt=0; else bin_j_opt=1;
   if prod_k=0 then bin_k=0; else bin_k=1;
   if prod_l=0 then bin_l=0; else bin_l=1;
   if prod_l_opt=0 then bin_l_opt=0; else bin_l_opt=1;
   if prod_m=0 then bin_m=0; else bin_m=1;
   if prod_n=0 then bin_n=0; else bin_n=1;
   if prod_o=0 then bin_o=0; else bin_o=1;
   if prod_o_opt=0 then bin_o_opt=0; else bin_o_opt=1;
   if prod_q=0 then bin_q=0; else bin_q=1;
run;
```

Now you should rerun the Merge node to complete the binary product data. When you view the results of the Merge node, look at the output data sets to view the data. We will now use a combination of a SAS macro and the Metadata node to complete our binary and quantity affinities by each segment. **Step 9:** Drag a Metadata node and connect the output of the Merge node to it. When you open the variables property sheet in the Metadata node, you should change the bin_ variable's entire new role to Prediction. **Step 10:** Now place another SAS Code node in the diagram and attach the Metadata node to it. I labeled this SAS Code node "Affinity by Segment." Figure 10.7 shows the SAS code computations for product quantities and binary product scores.

Figure 10.7 SAS Code for Affinity by Segment Computations

```
/* Calculating overall and cluster means   */
ods html style=barrettsblue body='c:\temp\bin_qty_means.htm';
title 'Product Quantity Affinity by Segment Means';
proc means data=EMWS.META_TRAIN mean ;
  class _segment_ ;
 var prod_a prod_b prod_c prod_d prod_e prod_f prod_g
     prod_h prod_i prod_j prod_k prod_l prod_m prod_n prod_o
     prod_p prod_q prod_a_opt prod_b_opt prod_e_opt prod_i_opt
     prod_j_opt prod_o_opt prod_l_opt ;
run;

title 'Product Binary Affinity by Segment Means';
proc means data=EMWS.META_TRAIN mean ;
  class _segment_ ;
 var bin_a bin_b bin_c bin_d bin_e bin_f bin_g
     bin_h bin_i bin_j bin_k bin_l bin_m bin_n bin_o
     bin_p bin_q bin_a_opt bin_b_opt bin_e_opt bin_i_opt
     bin_j_opt bin_o_opt bin_l_opt ;
run;
ods html close;
```

If you right-click on the Affinity by Segment SAS Code node, you can view the output results or the HTML output written to the relatively generic location of **`c:\temp\`**. If you want to manipulate these means by segment further, you can place an output statement in PROC SUMMARY, which would also output a SAS data set of the means by each segment. Figures 10.8 and 10.9 show the *partial* outputs for the product quantity and binary affinity scores, respectively.

Figure 10.8 Product Quantity Affinity by Segment

```
Output

Product Quantity Affinity by Segment Means

The MEANS Procedure

  _SEGMENT_          N Obs    Variable              Mean
----------------------------------------------------------
          1          37444    Prod_A           7.5665091
                              Prod_B          27.7325727
                              Prod_C           9.0002671
                              Prod_D          21.0512163
                              Prod_E           2.5094874
                              Prod_F          15.2284171
                              Prod_G          16.3641179
                              Prod_H          76.8762961
                              Prod_I           2.1539881
                              Prod_J           5.9286351
                              Prod_K           8.6394892
                              Prod_L          14.3246869
                              Prod_M           5.6988810
                              Prod_N          12.7297297
                              Prod_O           8.1889610
                              Prod_P           4.4248457
                              Prod_Q          24.6770300
```

Figure 10.9 Product Binary Affinity by Segment

```
Output


Product Binary Affinity by Segment Means

The MEANS Procedure

  _SEGMENT_          N Obs    Variable              Mean
----------------------------------------------------------
          1          37444    bin_a            0.0369886
                              bin_b            0.3963786
                              bin_c            0.9999733
                              bin_d            0.7458872
                              bin_e            0.0931524
                              bin_f            0.3743724
                              bin_g            0.4324591
                              bin_h            0.1070666
                              bin_i            0.3872717
                              bin_j            0.0295641
                              bin_k            0.0239291
                              bin_l            0.0286828
                              bin_m            0.0257184
                              bin_n          0.000908023
                              bin_o            0.0399797
```

Now, if you compare each segment's product quantity means with the other segments' means, you can tell if, on average, a segment has a higher or lower affinity for a particular product or set of products. Let's take Product B, for example. In the output of Figures 10.8 and 10.9, the five segments have Product B's means listed as shown in Table 10.1. It is clear in Table 10.1 that segment 3 has the highest product affinity for Product B, whereas segment 5 has the lowest. You can also see this clearly with the binary affinity scores as well. At the end of this chapter,

Exercise 1 gives you a chance to reformat the output of these two summary procedure steps so that the comparison of each product can be made easily.

Table 10.1 Comparison of Product B: Quantity vs. Binary Score Means by Segment

Segment	Product B's Average Quantity	Product B's Binary Mean Score
1	27.73	0.3964
2	65.26	0.6784
3	159.28	0.7460
4	85.63	0.4942
5	69.91	0.5422

The main results of analyzing these affinity scores is to compare the scores of each segment with each other in order to determine what set of products could up-sell or cross-sell based on the segment's averages. Now, campaigns can be fashioned for each segment using this data and tested against a differing product offer so that the effect of each campaign can be measured. Figure 10.10 shows the process flow diagram up to this point in this project.

Figure 10.10 Product Affinity Process Flow

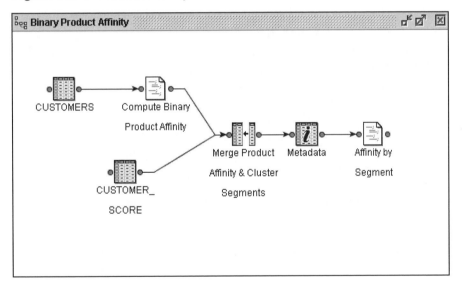

10.4 Pros and Cons of Segment Affinity Scores

The technique described in Section 10.3 indicates that five customer segments can be compared and contrasted using only the simple average. Although it is true that the average is considered as the central tendency theorem dictates, when the distribution is not a *normal* distribution as shown in Figure 10.2, the average may not truly represent the general central tendency for the group in which the average was computed. In other words, the average is not a good estimate of central tendency when the distribution is highly skewed. Also, in the case of the binary mean scores, when a customer purchases more than one item of a product, the binary score does not take this into account when being coded into 1s and 0s. Of course, we could take it into account so that if a customer purchases more than two or three of the same product, then the binary score gets a weight using another column to indicate a higher weighting than customers who purchase only a single item. Other weighting schemes could also be employed as well.

Even though these fallacies exist, they do offer some advantages as segment scores. For the binary scores, each product can be compared to other products by segment, whereas when quantities are used, each product has a differing distribution and high and low value; however, with the binary approach, every product is scaled from 0 to 1. This feature enables a comparison of one product to another since they all fall on a similar scaling system. One possible method for checking if the binary score approach is a good one is to see how often customers purchase more than one of the same product in the time period of the data set. If this does not happen very often, then perhaps the binary scaling is a rather good indicator; however, if it happens a lot, then a differing scaling method might be in order.

If you're getting the feeling that there is no one single technique that is best suited for understanding product affinity, then you're right, there isn't. However, when performing a technique such as this, one should understand the phrase that I believe Dr. G. E. P. Box, a well-known industrial statistician who published a great deal on experimental methods, coined "All models are wrong, but some models are useful." So, even if things are not as perfect in using quantities or binary affinity scores, they still may serve our purpose well enough to get the task at hand complete with satisfactory results. With this in mind, let's move on to the next topic to see if we can cluster the quantity or binary affinity scores.

10.5 Issues with Clustering Non-normal Quantities

As Figure 10.2 showed, the typical distribution of product quantities on the CUSTOMER data set is highly non-normal. This non-normality will greatly affect clustering algorithms in that highly skewed distributions will cause the algorithm to place the skewed items in their own separate cluster, when that may not be the best representation when clustering customer data. Let's see what happens to product quantity data in a cluster situation without any normalization.

Step 11: In the Product Affinity process flow diagram, drag a Cluster node onto the flow diagram and connect the Metadata node to it. In the Cluster node property sheet, open the Variables icon and select only the Cust_ID field (mandatory for clustering to contain a single ID field) and set the Use column to Yes only for Prod_A, Prod_B, and Prod_C fields. Set the Use column to No for all other fields. Let the other clustering property settings to be at their default settings. Now run the Cluster node. What you should obtain is a set of around 22 clusters with a wide range of numbers of customers per cluster. Many of the clusters contain only a single customer. This is probably an outlier that the algorithm detected and gave it its own cluster membership. Figure 10.11 shows some of the cluster statistics and numbers within each cluster, and Figure 10.12 shows the cluster distance table.

Figure 10.11 Product A, B, and C Quantity Cluster Statistics

SEGMENT...	Frequency of Cluster	Root-Mean-Square St...	Maximum D...	Near
1	92	361.8698	1283.156	
2	1	.	0	
3	2474	146.1271	1812.83	
4	1	.	0	
5	32	522.685	1726.283	
6	1	.	0	
7	372	278.8038	1102.274	
8	1	.	0	
9	1	.	0	
10	2	352.3043	431.4829	
11	2	264.8282	324.347	
12	29	569.0351	2784.846	
13	102006	28.16672	1477.354	
14	411	208.824	1256.191	
15	1	.	0	
16	1	.	0	
17	1	.	0	
18	2	264.4705	323.9089	
19	8	560.5509	1816.184	
20	6	263.0328	562.0492	
21	2	627.2385	768.2072	
22	19	413.4087	1168.609	

Figure 10.12 Product A, B, and C Quantity Cluster Distance Plot

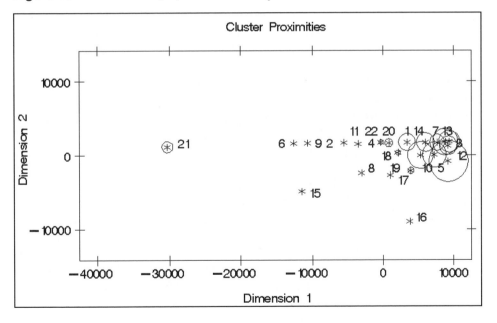

What we see in the distance plot is that many clusters are so close in distance, that the algorithm did not do an effective job of cluster membership or separation due to the fact that these product quantity distributions are highly non-normal and skewed. Algorithms have been developed to cluster data with properties like this; however, most have not made it to commercially available software packages such as SAS. **Step 12:** Now, to see the effect on the same set of products we just clustered, copy the Cluster node and paste a copy just below the first one.

Step 13: Now drag a Transform Variables node, connect the Metadata node to it, and then connect the Transform Variables node to the second Cluster node as shown in Figure 10.13.

Figure 10.13 Transformed Product A, B, and C Quantities Clustering

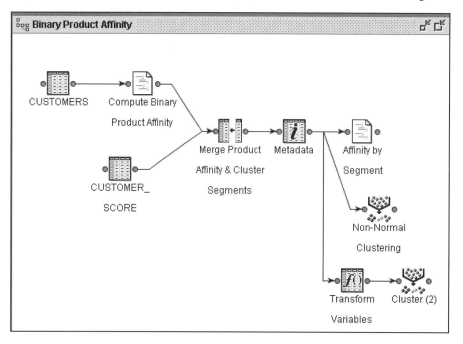

Open the Variables icon in the Transform Variables node property sheet and set Prod_A, Prod_B, and Prod_C to Max Normal and all other variable to None. This will use an algorithm to determine the best set of possible transformations in order to maximize normality of the quantity data for these three variables. **Step 14:** Now, if we just run the second Cluster node, then we should obtain a better clustering than we did in the non-normal case. However, if we set the Internal Standardization to Range and set the Clustering Method to Centroid rather than the default Ward method. So, let's rerun this Cluster node with these settings to see if we can improve the product clusters. Now, look at the distribution in the number of customers per cluster (view the Mean Statistics table for this) and also the Cluster plot shown in Figure 10.14. There are still many outliers and clusters with only a single customer. We seem to be getting better; however, this may not be a usable customer clustering as it currently stands. **Step 15:** Now, let's try something else. Instead of letting the system estimate the number of clusters, set the Specification Method to "User Specify" and the Maximum to 12 instead of the default. Keep the current settings for Range standardization and the Centroid method. Rerun the Cluster node and see what this does to our cluster analysis. If you look at the Cluster distance plot shown in Figure 10.14, what we see now is concentric circles of clusters when we have forced the number. This also may not be very desirable as several clusters are entirely overlapping, and this will not allow a separate and distinct segmentation of customers with these three products.

As you might have guessed, clustering of product data is a bit more complex than what we have attempted up to this point. We either have to modify our transformations in order to make the distributions even closer to normality, or perhaps modify the algorithm(s) we use to group customers into like segments, or perhaps even both. In Chapter 6, "Clustering of Many Attributes," I introduced briefly the softmax function in Equation 6.1. This function has the properties of scaling data between 0 and 1. Perhaps, if we scale the three product quantities Prod_A, Prod_B, and Prod_C using the softmax function, then this may aid in clustering the product data. **Step 16:** Place a SAS Code node on your process flow diagram and attach the Metadata node to it. In the SAS code, place the code of the softmax.sas file from the CD and the three macro calls for each of the products.

Figure 10.14 Cluster Plot of Normalized Products A through C with Range Standard and Centroid Method

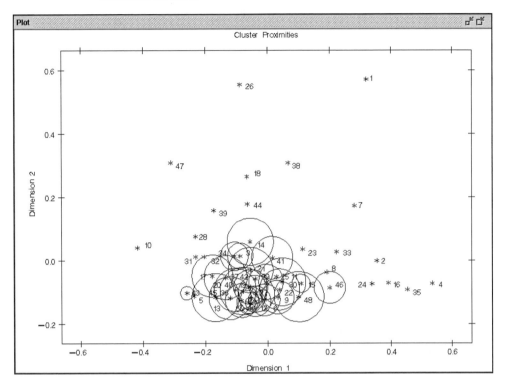

Figure 10.15 Cluster Plot of Normalized Products A through C with Range and Centroid and Fixed Number of Clusters Set to 12

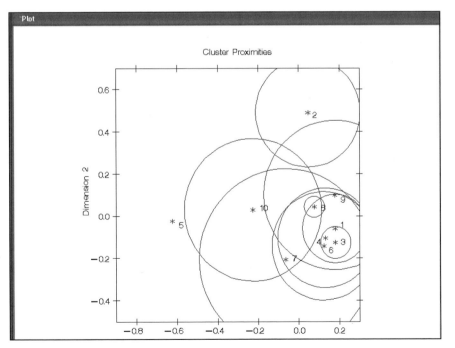

Figure 10.16 shows the SAS code for the node. There is also a softmax call routine function in the Base SAS function library. We will compare the macro version and the SAS function version of the softmax computations as well. The scored data set will be saved in the SAMPSIO data library, which we will then add into the Data source folder.

Figure 10.16 SAS Code for Softmax Transformations

```
%macro softmax(dsin=,var=,dsout=,log=L);
 proc sql;
   create table work.stats as
  select min(&var) as minv,
         max(&var) as maxv,
         mean(&var) as meanv,
         std(&var) as stdev
    from &dsin
 ;
quit;
   data _null_ ;
     set work.stats;
       call symput('minv',minv);
       call symput('maxv',maxv);
       call symput('meanv',meanv);
       call symput('stdev',stdev);
   run;
%if %upcase(&log)=L %then
     %do;
        data &dsout;
        set &dsin;
         sm_&var = (&var - &minv)/(&maxv - &minv);
        run;
     %end;
 %else
   %if %upcase(&log)=S %then
       %do;
          data &dsout;
             set &dsin;
             %let var1 = (&var - &meanv)/(1.283 *
(&stdev/6.2831853));
             sm_&var = 1/(1 + exp(- &var1));
          run;
       %end;
 %else
   %if %upcase(&log)=SS %then
       %do;
          data &dsout;
             set &dsin;
              %let var1 = (&var - &meanv)/(1.283 *
(&stdev/6.2831853));
             %let var2 = 1/(1 + exp(- &var1));
                sm_&var = &meanv +
                ((1.283 * &stdev*log(sqrt(-1+(1/(1-&var2)))) )/
3.14159265 );
             run;
       %end;
%mend softmax;
%softmax(dsin=EMWS.META_TRAIN,dsout=work.temp_a,var=prod_a,log=S);
%softmax(dsin=work.temp_a,dsout=work.temp_b,var=prod_b,log=S);
%softmax(dsin=work.temp_b,dsout=sampsio.soft_score,var=prod_c,log=S);
```

Step 17: Now after the three softmax macro calls in Figure 10.17, place SAS code shown in Figure 10.17. The entire code in Figures 10.16 and 10.17 is in Appendix 3 in the data CD under Chapter 10 in the file softmax_transform.sas. The code in Figure 10.17 rewrites the same data set SAMPSIO.SOFT_SCORE with three new variables SAS_SOFT_PRODA, SAS_SOFT_PRODB, and SAS_SOFT_PRODC. The call to the function CALL SOFTMAX computes the softmax for each of the product affinities A through C and computes this for each customer ID. This is necessary as SAS is computing the function call on each customer, whereas the macro routine does this by using a do-end section, as shown in Figure 10.16.

Figure 10.17 SAS Code for Softmax Transformations, Continued

```
/* Now compare the Softmax macro to SAS softmax function */
proc sort data=sampsio.soft_score;
  by cust_id;
run;
data sampsio.soft_score;
   set sampsio.soft_score;
sas_soft_proda = prod_a;
sas_soft_prodb = prod_b;
sas_soft_prodc = prod_c;
call softmax(sas_soft_proda,sas_soft_prodb,sas_soft_prodc);
  by cust_id;
run;
quit;
```

Recall that in Chapter 6, Equation 6.1 contains two components for the data scaling; in the first part, the e^{-Xt} portion scales the data with a logistic function (the data squashing part), and the $X_t = \dfrac{(X_i - \bar{X})}{\lambda(\sigma_i / 2\pi)}$ part does the standardizing against the mean and standard deviation. The SAS call routine CALL SOFTMAX has a slightly different form. From the SAS 9.1 documentation (SAS Institute 2003b), the SAS call routine computes the softmax function in a slightly different manner. The equation that SAS uses to compute softmax is as follows:

$$X_j = \frac{e^{-Xj}}{\sum\limits_{i=1}^{i=n} e^{-Xi}}$$

(10.1)

This equation does differ from Equation 6.1. **Step 18:** Once your SAS Code node contains all of the code listed in Figures 10.16 and 10.17, you can run the SAS Code node and add the new data set SOFT_SCORE that is in the SAMPSIO SAS library to the Data sources folder. **Step 19:** Add two more Cluster nodes to your diagram. In the first Cluster node, we will call this node Clusters of Softmax where we will use the newly transformed Prod_A, Prod_B, and Prod_C from our macro computations, and these new variables are added to the data set as SM_PROD_A, SM_PROD_B, and SM_PROD_C respectively. In this Cluster node, set the Internal Standardization of the property sheet to Range, the Cluster Method to Average, and the maximum number of clusters to 10. Edit the variables in this Cluster node to use only the variables SM_PROD_A, SM_PROD_B, and SM_PROD_C. In the second Cluster node, which we'll call Clusters of SAS Softmax, set the Internal Standardization to Standardization, the Cluster Method to Average, and the maximum number of clusters to 10. Use only the variables SAS_SOFT_PRODA, SAS_SOFT_PRODB, and SAS_SOFT_PRODC. Now your process flow diagram should look like that in Figure 10.18. In both of these Cluster nodes, be sure to keep the

CUST_ID variable as that is necessary for each of the Clustering nodes to work properly. Now run both of the new Cluster nodes.

Figure 10.18 Revised Process Flow Diagram with Softmax Scaling of Products A through C

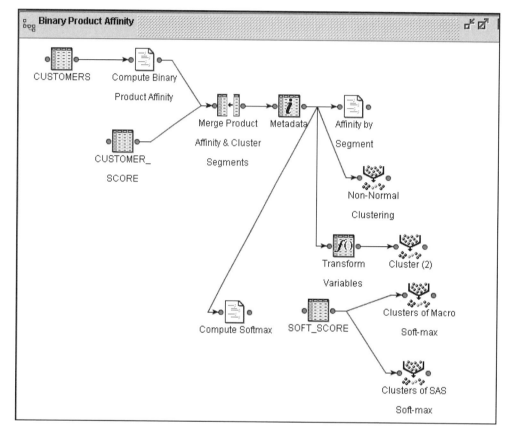

You should notice that these produce very different results of clusters than when we attempted to transform the product quantities earlier as shown in Figures 10.14 and 10.15. It appears that the macro-computed softmax product affinities are somewhat better cluster solutions than the SAS softmax computations, as the separation of cluster centers is a bit more uniform. Each of these softmax-scaled product affinities is much better, however, than we computed earlier. As a separate exercise, you should attempt to do Exercise 2 at the end of the chapter, which computes the softmax for all product affinities and clustering for all those scaled products.

So, with these techniques we've seen that if the business problem at hand needs product affinities for other customer segments, then overlaying the affinities onto segments shown earlier is appropriate; however, if the business need is to have customer segments defined by their product affinity, then the method of scaling those purchase quantities and clustering is a good choice. The technique should match the business problem at hand and therefore aid the business with customer analytics.

10.6 Approximating a Graph-Theoretic Approach Using a Decision Tree

One way to approach the clustering of product affinity data, or any other data which is highly non-normal, skewed, or otherwise ugly in nature, is to turn the clustering problem into a decision tree problem. In Chapter 4, "Segmentation Using a Cell-based Approach," a decision tree model segmented customer data using several attributes, including RFM cells and the like. Each final leaf of the decision tree is in fact a cluster that has the purity of that branch's variable as its similarity measure. The more records with the same leaf characteristics, the more *pure* the statistic called the Gini Index became. In essence, a decision tree algorithm, which uses a Gini Index to measure leaf purity, is somewhat analogous to a clustering algorithm (Duda, Hart, and Stork 2001, pp. 566–567). One of the main benefits of using a decision tree for clustering is as follows:

- Computationally efficiency is especially attractive for large data sets and many variables.

- The decision tree algorithm does not depend on the detailed geometric shape of the clusters.

- The outcome of the decision tree is not dependent on the distributional properties of the input variables that feed the decision tree algorithm.

This makes a decision tree very attractive for the purposes of clustering and segmenting our data. This technique has been applied to very large industrial data sets where there are many variables and the desire to understand the functional relationships in the data is important. The application of this technique in biological gene-expression data has been recently documented in the literature (Y. Xu, Olman, and D. Xu 2002, pp. 536–545). The reasons for using graph-theoretic approaches are similar to those in CRM applications.

Graph theory has been applied to problems where there is very high dimensionality. In general, the more variables and diversity of those variables there are on a data set, the greater the dimensionality. The general principle for clustering with a graph is like connecting the dots for all the data points, then cutting the lines between the longest lines where the group of dots has the most similarity. Figure 10.19 shows that the removal of inconsistent edges (ones with length significantly longer than the average) might yield natural clusters (Duda, Hart, and Stork 2001, pp. 566–567).

Figure 10.19 Example of Graph Theory Finding Data Clusters

(Richard O. Duda, Peter E. Hart, and David G. Stork, Pattern Classification. Copyright © 2001 by John Wiley & Sons, Inc.)

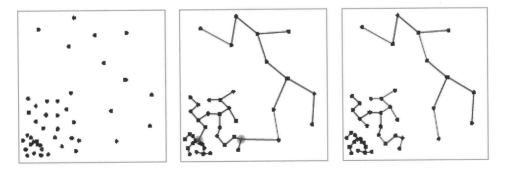

When we used a decision tree before to predict segments, we used any set of variables on the data set, which were appropriate to predict the target variable, in this case a segment level. In our case now, we are not looking to predict the segment; however, we are looking at the resulting set of leaves at the bottom of the tree. These leaves are the clusters. We will use only the product quantities to estimate the customer segment variable EM_SEGMENT as our target, and we'll first start with the three product quantity scores Prod_A, Prod_B, and Prod_C as inputs to the decision tree. In order for the decision tree to approximate the clustering problem, a Minimum Spanning Tree (MST) is needed. A key property of the MST is that each cluster of the product affinity data corresponds to one sub-tree of the MST, which rigorously converts a multidimensional clustering problem to a decision tree partitioning problem. The minimum spanning tree using Euclidian distances (EMST) is when the EMST algorithm connects a set of dots using lines such that the total length of all the lines is minimized and any dot can be reached from any other by following the lines. This is much like a more difficult version of the child's game connect-the-dots. Although I won't bore you with all of the rigorous mathematical proofs here, a picture of how this problem transcends from one algorithm to another will be useful. If you would like to know more about the mathematics of the MST and clustering, Y. Xu, Ohnan, and D. Xu (2002, pp. 536–545) discuss the proof of converting a multidimensional clustering problem into a decision tree problem.

Process Flow Table 2: Graph-Theory Approach

Step	Process Step Description	Brief Rationale
1	Create a new process flow diagram called Graph-Theory Approach.	
2	Add the SOFT_SCORE data from Data Sources folder to the diagram.	Sets the _SEGMENT_ variable as the target.
3	Add a Data Partition node and use the default settings *except set stratified.*	Splits data into training, validation, and test.
4	Drag a Decision Tree node to the diagram and connect the Data Partition to it.	Uses a decision tree model to estimate product affinity segments.
5	Drag a Segment Profile node and use default settings.	Our interest is the _NODE_ variable.
6	Add a SAS Code node and connect the Segment Profile node to it.	Computes mean softmax score by segment.

Step 1: Let's build a decision tree model from our three product quantities using the concepts of clustering and the MST. We'll use the data set SOFT_SCORE that we created in our last diagram. Create a new process flow diagram and call it Graph-Theory Approach. **Step 2:** Drag the SOFT_SCORE data set from the Data Sources folder onto the flow diagram space. The cluster segment we created before called _SEGMENT_ will be our target variable, and the three product affinities will be the quantities of Prod_A, Prod_B, and Prod_C. Be sure to set the _SEGMENT_ as your target variable. **Step 3:** Also, drag a Data Partition node and connect the SOFT_SCORE data to it. The default settings for training, validation, and test set sample sizes are fine to use as they are. Open Variables in the property sheet and set the _SEGMENT_ target variable's partition role to Stratified. **Step 4:** Now, drag a Decision Tree node to the diagram and connect the Data Partition node to the Decision Tree node. We will attempt approximate an MST using the Decision Tree property sheet settings shown in Figure 10.20.

Figure 10.20 Decision Tree Property Settings to Approximate an MST Algorithm

Property	Value
Criterion	Gini
Significance Level	0.2
Missing Values	Use in search
Use Input Once	Yes
Maximum Branch	3
Maximum Depth	6
Minimum Categorical S	5
⊟Node	
Leaf Size	5
Number of Rules	5
Number of Surrogate R	0
Split Size	
⊟Split Search	
Exhaustive	5000
Node Sample	5000
⊟Subtree	
Method	Assessment
Number of Leaves	1
Assessment Measure	Misclassification
Assessment Fraction	0.25
⊟P-Value Adjustment	

The Method property (under the Subtree section) should be set to Assessment. This will produce the smallest subtree with the best assessment value. In addition, Maximum Branch should be set to 3 (to allow more than just a binary branch tree), and the Criterion should be set to Gini, which will measure the impurity of each product group at each node (and will try to maximize the purity). The Use Input Once property should also be set to Yes as that way a subtree will not consist of multiple combinations of products. These settings should come closest to approximating the MST algorithm; however, it will not be an exact representation. Now run the Decision Tree node.

The results of the Decision Tree will produce a prediction for each _SEGMENT_ level; however, we won't be looking at those here as those are the predicted values of each segment level. We want to look at the variable called _NODE_. This variable shows the node number used in each decision. Notice that there are seven distinct nodes used in this Decision Tree model. Figure 10.21 shows the actual tree model and Figure 10.22 shows the decision rules for each node.

Figure 10.21 Decision Tree Model Results (Tree Diagram)

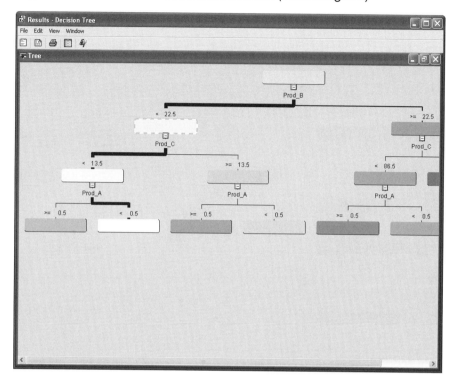

The decision rules of Figure 10.22 indicate that there are six decisions at each node. We will want to group customers by each of these decisions rather than use the predictive levels of the _SEGMENT_ variable. We will also want to profile the customer data by using the Segment Profiler node. **Step 5:** Drag a Segment Profile node and connect the Decision Tree node to the Profile Node. Now run the Profile Node. The _NODE_ variable is considered to be a segment variable to the Profile Node by default because the _NODE_ variable role is a segment.

Step 6: Now place a SAS Code node at the end of Segment Profile node and connect the Segment Profile to it. The following SAS code should compute the mean values for each _NODE_ level for the softmax scaled products following A through C, respectively.

Figure 10.22 SAS Code Node PROC MEANS Statement

```
Proc means data=&EM_IMPORT_DATA mean ;
Class _node_ ;
 Var sm_prod_a sm_prod_b sm_prod_c;
Run;
```

Figure 10.23 Decision Tree Model (Tree Rules and Nodes)

```
IF          86.5 <= Prod_C          IF          0.5 <= Prod_A
AND         22.5 <= Prod_B          AND        13.5 <= Prod_C
THEN                                AND Prod_B <          22.5
   NODE    :       7                THEN
   N       :     692                   NODE    :      11
   AVE     : 2.97832                   N       :     400
   SD      : 1.02334                   AVE     : 2.7675
                                       SD      : 1.2363
IF  Prod_A  <         0.5
AND Prod_C  <        13.5            IF  Prod_A  <         0.5
AND Prod_B  <        22.5           AND Prod_C  <        86.5
THEN                                AND        22.5 <= Prod_B
   NODE    :       8                THEN
   N       :   28120                   NODE    :      12
   AVE     : 2.30957                   N       :    5776
   SD      : 1.47229                   AVE     : 2.72836
                                       SD      : 1.17059
IF          0.5 <= Prod_A
AND Prod_C  <        13.5            IF          0.5 <= Prod_A
AND Prod_B  <        22.5           AND Prod_C  <        86.5
THEN                                AND        22.5 <= Prod_B
   NODE    :       9                THEN
   N       :    1417                   NODE    :      13
   AVE     : 2.64926                   N       :    1957
   SD      : 1.26922                   AVE     : 2.87021
                                       SD      : 1.05472
IF  Prod_A  <         0.5
AND         13.5 <= Prod_C
AND Prod_B  <        22.5
THEN
   NODE    :      10
   N       :    3733
   AVE     : 2.4996
   SD      : 1.30663
```

What the Decision Tree model has done for us is to group product affinities of A, B, and C by using a minimum spanning tree (approximated). This is yet another form of clustering; however, we did not perform a clustering algorithm on our customer data. The use of a decision tree has enabled us to cluster customers together with similar product affinities. Figure 10.24 shows the mean scores for each cluster node from the Decision Tree, and the final completed process flow diagram is shown in Figure 10.25.

Figure 10.24 SAS Code Node Output of Product Affinity Mean Scores by Decision
Tree Node

```
Product Binary Affinity by Segment Means

The MEANS Procedure

        Node        N Obs    Variable          Mean
  -----------------------------------------------------------
           7          693    sm_prod_a     0.5405876
                             sm_prod_b     0.8316820
                             sm_prod_c     0.9968462

           8        28191    sm_prod_a     0.4055003
                             sm_prod_b     0.3417277
                             sm_prod_c     0.3424678

           9         1417    sm_prod_a     0.6066076
                             sm_prod_b     0.3515061
                             sm_prod_c     0.3488271

          10         3744    sm_prod_a     0.4055028
                             sm_prod_b     0.3441345
                             sm_prod_c     0.6997059

          11          401    sm_prod_a     0.6337901
                             sm_prod_b     0.3547558
                             sm_prod_c     0.7303992

          12         5780    sm_prod_a     0.4054908
                             sm_prod_b     0.6287582
                             sm_prod_c     0.4488536

          13         1960    sm_prod_a     0.7200858
                             sm_prod_b     0.7350540
                             sm_prod_c     0.4687073
  -----------------------------------------------------------
```

Figure 10.25 Decision Tree Clustering Process Flow Diagram

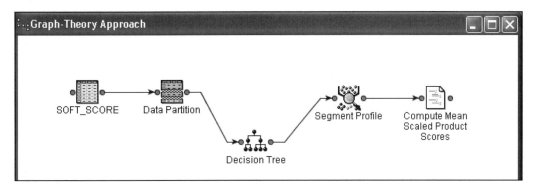

10.7 Using the Product Affinities for Cross-Sell Programs

Product affinity scores can be used to aid in the analysis of what a customer's product profile looks like. An analyst armed with this knowledge can use this information to help design a product or service cross-sell program or offering. In the additional exercises at the end of this chapter, Exercise 1 asks you to reformat the average product affinity scores by each segment. Figure 10.26 shows what this output should look like. In Figure 10.26, say Prod_L is of interest to a product marketing manager. The overall binary mean score for all segments is 0.0908. Segment 3 has the largest affinity for this product at 0.22156; however, segments 4 and 5 are rather low (0.036 and 0.016, respectively). To cross-sell Prod_L, the marketing manager should focus that product's offering into segments 4 and 5. Once the customers are profiled in those segments, separate messaging should also accompany the cross-sell offerings according to the profile of segments 4 and 5. Many other product cross-sell opportunities exist in this data and an entire program of campaigns could be launched just doing a number of cross-sell product offerings.

Now, this example is somewhat simplistic as it currently stands, since segments 4 and 5 have the lowest product affinity scores of most or all of the products and options. So, other characteristics like the amount segments 4 and 5 are likely to spend on a suite of products might assist in understanding why their overall affinity scores are rather low in caparison to the other segments. In Figure 10.27 the results of the Segment profile node is shown which allows more in-depth comparisons of each of these segments.

Figure 10.26 Re-Formatted Product Affinity Scores by Segment (from Section 10.3)

	NAME OF FORMER VARIABLE	COL1	COL2	COL3	COL4	COL5	COL6
1	_SEGMENT_	.	1	2	3	4	5
2	bin_a	0.096752477	0.128460009	0.051284221	0.216929209	0.038887447	0.0361826759
3	bin_b	0.549973925	0.602385686	0.542377184	0.793812868	0.456700407	0.3025244933
4	bin_c	0.999962073	1	0.999915373	1	1	0.9998963247
5	bin_d	0.841701038	0.884768313	0.855117844	0.965913836	0.814776365	0.6587009486
6	bin_e	0.113573223	0.215935158	0.006600939	0.177263238	0.082360066	0.1256544503
7	bin_f	0.523187787	0.686496406	0.393940676	0.793926741	0.396444329	0.3529625214
8	bin_g	0.563542407	0.667839119	0.495705158	0.785689884	0.501254434	0.3472085428
9	bin_h	0.15474328	0.239103839	0.062116532	0.261226039	0.124621507	0.1017054585
10	bin_i	0.501322714	0.432940817	0.591968857	0.659897514	0.451250108	0.2800787932
11	bin_j	0.04223202	0.078375898	0.001523294	0.074663124	0.026256597	0.0424550308
12	bin_k	0.112549187	0.098409543	0.075614607	0.29982919	0.016523921	0.0266963869
13	bin_l	0.09087375	0.099097721	0.054880887	0.221560068	0.03624881	0.0163806957
14	bin_m	0.040989902	0.077076006	0.000634706	0.078762574	0.025737521	0.032657716
15	bin_n	0.008780164	0.005964215	0.005543097	0.025849307	0.000605589	0.0011404282
16	bin_o	0.062238657	0.120584187	0.001650235	0.116302904	0.036292067	0.0541703385
17	bin_p	0.361845162	0.396696743	0.292006939	0.653141013	0.220823601	0.194961381
18	bin_q	0.517280614	0.636718153	0.41585918	0.799506548	0.37836318	0.3416100772
19	bin_a_opt	0.302441568	0.370928276	0.248593069	0.564015942	0.181849641	0.1092737546
20	bin_b_opt	0.788384772	0.855329561	0.781280413	0.951945341	0.729907431	0.5984137681
21	bin_e_opt	0.150960034	0.295381557	0.006685567	0.230480167	0.108573406	0.1719973044
22	bin_i_opt	0.007557009	0.003899679	0.00410443	0.023382046	0.000216282	0.001451454
23	bin_j_opt	0.105343005	0.201254014	0.006685567	0.16796356	0.069296652	0.1188637188
24	bin_o_opt	0.030664201	0.059106897	0.001692549	0.065401404	0.015745307	0.0173137733
25	bin_l_opt	0.071568767	0.069964826	0.040028773	0.186980452	0.021109092	0.0141516769

Figure 10.27 Partial Results of Segment Profile Node Output

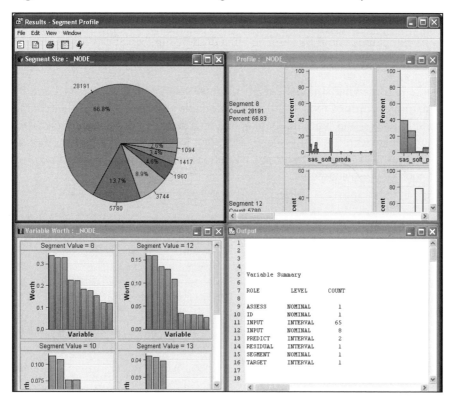

In all of these techniques, the overall idea is to gain a better understanding of the customer's needs, desires, purchase patterns, and revenue stream so that customers receive the most relevant marketing or sales Product affinity scoring allows the scaling of product ownership or even product portfolio ownership in a fashion so that each customer can be ranked on a scale from 0 to 1. The methods I've shown in this chapter involve simple ownership patterns, such as 0 if customer has no product and 1 if the customer has one or more of the product; however, more elaborate schemes can be used if needed. The next chapter discusses how a special purpose neural network called self-organizing map (SOM) can be used to cluster and segment customers or prospects into a two-dimensional map.

10.8 Additional Exercises

1. In Section 10.3, the product affinity scores (both quantity and binary) were output to a report that is somewhat difficult for comparison purposes. Reformat the report so that the scores appear as follows: Column 1 is the grand average, and columns 2 through 6 are the averages for segments 1 through 5, respectively. This format allows for a much easier visual comparison.

	NAME OF FORMER VARIABLE	COL1	COL2	COL3	COL4	COL5	COL6
1	_SEGMENT_	.	1	2	3	4	5
2	bin_a	0.096752477	0.128460009	0.051284221	0.216929209	0.038887447	0.0361826759
3	bin_b	0.549973925	0.602385686	0.542377184	0.793812868	0.456700407	0.3025244933
4	bin_c	0.999962073	1	0.999915373	1	1	0.9998963247
5	bin_d	0.841701038	0.884768313	0.855117844	0.965913836	0.814776365	0.6587009486
6	bin_e	0.113573223	0.215935158	0.006600939	0.177263238	0.082360066	0.1256544503
7	bin_f	0.523187787	0.686496406	0.393340676	0.793926741	0.396444329	0.3529625214
8	bin_g	0.563542407	0.667839119	0.495705158	0.785689884	0.501254434	0.3472085428
9	bin_h	0.15474328	0.239103839	0.062116532	0.261226039	0.124621507	0.1017054585
10	bin_i	0.501322714	0.432940817	0.591968857	0.659897514	0.451250108	0.2800787932
11	bin_j	0.04223202	0.078375898	0.001523294	0.074663124	0.026256597	0.0424550308
12	bin_k	0.112549187	0.098409543	0.075614607	0.29982919	0.016523921	0.0266963869
13	bin_l	0.09087375	0.099097721	0.054880887	0.221560068	0.03624881	0.0163806957
14	bin_m	0.040989902	0.077076006	0.000634706	0.078762574	0.025737521	0.032657716
15	bin_n	0.008780164	0.005964215	0.005543097	0.025849307	0.000605589	0.0011404282
16	bin_o	0.062238657	0.120584187	0.001650235	0.116302904	0.036292067	0.0541703385
17	bin_p	0.361845162	0.396696743	0.292006939	0.653141013	0.220823601	0.194961381
18	bin_q	0.517280614	0.636718153	0.41585918	0.799506548	0.37836318	0.3416100772
19	bin_a_opt	0.302441568	0.370928276	0.248593069	0.564015942	0.181849641	0.1092737546
20	bin_b_opt	0.788384772	0.855329561	0.781280413	0.951945341	0.729907431	0.5984137681
21	bin_e_opt	0.150960034	0.295381557	0.006685567	0.230480167	0.108573406	0.1719973044
22	bin_i_opt	0.007557009	0.003899679	0.00410443	0.023382046	0.000216282	0.001451454
23	bin_j_opt	0.105343005	0.201254014	0.006685567	0.16796356	0.069296652	0.1188637188
24	bin_o_opt	0.030664201	0.059106897	0.001692549	0.065401404	0.015745307	0.0173137733
25	bin_l_opt	0.071568767	0.069964826	0.040028773	0.186980452	0.021109092	0.0141516769

2. In Section 10.5, we computed scaled product affinity scores using the softmax macro function for three products PROD_A, PROD_B, and PROD_C. In this exercise, compute the softmax scaling for all of the product categories and options and cluster all of those to see what kind of cluster segments arise from the softmax scaled product quantity data.

10.9 References

Berry, Michael J. A., and Gordon S. Linoff. 2004. *Data Mining Techniques: for Marketing, Sales, and Customer Relationship Management.* 2d ed. New York: John Wiley & Sons, Inc.

Duda, Richard O., Peter E. Hart, and David G. Stork. 2001. *Pattern Classification.* 2d ed. New York: John Wiley & Sons, Inc.

SAS Institute Inc. 2003a. "Association Node Reference." Enterprise Miner, Release 5.2. Cary, NC: SAS Institute Inc.

SAS Institute Inc. 2003b. "Base SAS Functions and Call Routines." *SAS 9.1 Documentation.* Cary, NC: SAS Institute Inc.

Xu, Ying, Victor Olman, and Dong Xu. 2002. "Clustering gene expression data using a graph-theoretic approach: an application of minimum spanning trees." *Bioinformatics.* 18.4:536–545.

Computing Segments Using SOM/Kohonen for Clustering

11.1 When Ordinary Clustering Does Not Produce Desired Results

In many cases, my experience has found that transforming numeric variables and combining character data typically is sufficient to use the clustering techniques we have reviewed in the last several chapters. At times, however, there are certain problems where transforming the data still does seem to produce satisfactory clusters. By satisfactory, I mean that there is not a clear distinction in the cluster segments among the attributes selected for the clustering process. The clusters may contain attributes, which span several or even most clusters and therefore don't seem to consolidate into one or two more distinct and unique clusters. There are times that the clustering itself does not produce the desired number of clusters and will produce only two clusters and consequently is not very useful. When issues like these come about, an alternate technique is sometimes valuable for performing clustering.

11.2 What Is a Self-Organizing Map?

Self-organizing maps (SOMs) are an unsupervised data visualization technique, invented by Professor Teuvo Kohonen (1981; 1988), which reduce the dimensions of data through the use of self-organizing neural networks. Although this doesn't sound very useful, a neural network is actually a computer model of how biological neurons interconnect and operate. If you can picture a neuron like that of Figure 11.1, and imagine a human brain containing around 10 billion of these interconnected neurons, then a neural network is a model of how these biological components work. A neuron will *fire* when some threshold is met, which might be something like heat coming in contact with your skin. These neurons will send a signal from your finger to your brain, and you compile these inputs and realize that your finger is hot. Your brain sends a signal to the muscles in your arm and hand to retract away from the heat source.

Figure 11.1 Biological Schematic of Neuron

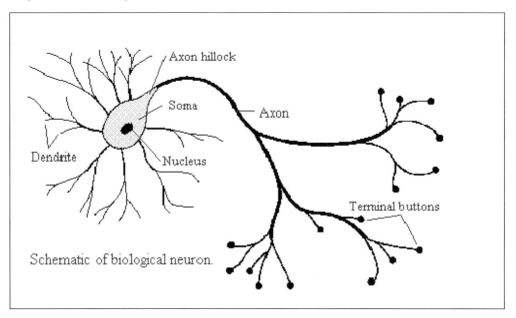

This simple example is rather similar to a computer-generated model of this neuron. It was thought that simulating a biological system where a great many inputs are very successfully processed communicated to our brains could also be valuable as a computer-generated algorithm.

As complicated as the biological neuron is, it may be simulated by a very simple model like Figure 11.2. The inputs each have a weight that they contribute to the neuron, if the input is active. The neuron can have any number of inputs; neurons in the brain can have as many as a thousand inputs. Each neuron also has a threshold value. If the sum of all the weights of all active inputs is greater than the threshold, then the neuron is active. For example, consider the case where both inputs are active. The sum of the input's weights is 0. Since 0 is smaller than 0.5, the neuron is off. The only condition that would activate this neuron is if the top input were active and the bottom one were inactive. This single neuron and its input weighting perform the logical expression *A and not B*, which is represented by the circle at the far right.

Figure 11.2 Computer Rendition of a Neuron

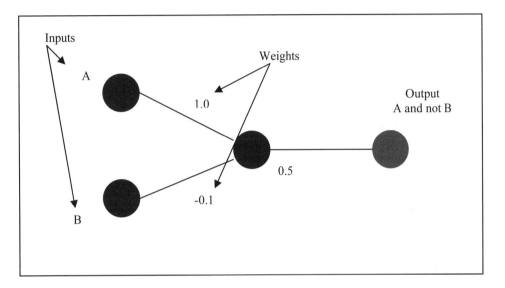

Now, back to SOM networks; suppose you were to select a group of attributes and variables from your data set, and create an interconnect network with the special condition for placing each row in your table into a two-dimensional map. SOMs are a data visualization technique, which reduces the dimensions of data through the use of self-organizing neural networks. The problem that data visualization attempts to solve is that humans simply cannot visualize high dimensional data as is, so techniques are created to help us understand this high dimensional data. So SOM networks accomplish two things: they reduce dimensions and they display similarities of those dimensions. As far as clustering and segmenting customers into distinct like groups then, this technique is fairly ideal (in principle anyway) for CRM applications. A way to visualize this in a simplistic example might be to say you have a data set where retail produce customers purchased items from a grocery store. You would like to group the customers according to the similarity of items purchased so that customers who purchased a similar set of items would be grouped together. A two-dimensional SOM segmentation might look like that shown in Figure 11.3. As you can see, each group has a label for the most typical items purchased in their basket.

Figure 11.3 SOM of Grocery Customer Purchases

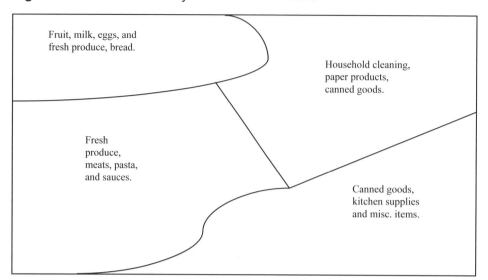

Although the map in Figure 11.3 is somewhat simplistic in nature, the main theme is that customers with similar purchase items are grouped together. These groups can then be used for marketing and sales efforts mentioned in earlier chapters. Some of the advantages and disadvantages of SOM are as follows:

Advantages

- They are conceptually easy to understand.

- SOMs tend to work very well.

- In SAS Enterprise Miner, the profiling portion is very similar to the clustering technique (more on this later in this chapter).

Disadvantages

- SOM networks can be prone to issues with missing data as in all other neural network algorithms and regressions.

- SOMs can produce differing results as they produce maps from sampled data so it may take a number of trials to obtain a map that is consistent with the same training data.

- They are rather computationally intensive.

- They have no good method for analyzing shelf-life of segments as we did earlier in Chapter 7, "When and How to Update Your Cluster Segments."

11.3 Computing and Applying SOM Network Cluster Segments

There are currently three options in SAS Enterprise Miner Release 5.2 that fit different types of neural network models. As the name of the SOM/Kohonen node suggests, two methods are implied: SOM and Kohonen. There are three ways to perform SOM and vector quantization (VQ) in SAS Enterprise Miner. In the SOM/Kohonen node, the method property has three levels as shown in Figure 11.4: Batch SOM, Kohonen SOM, and Kohonen VQ. The batch mode runs in the background and is the default setting. This mode should be used when the data set size is large with many rows and many variables. The size of the data set and the type of computer you are running SAS Enterprise Miner on might depend on when you want to run in batch or not. For our data set of 100,000 or so records and the number of variables, the SOM/Kohonen or VQ will run in less than a minute on most Windows-based servers.

Figure 11.4 Methods in the SOM/Kohonen Node Property Sheet

Property	Value	
Node ID	SOM	^
Imported Data	[...]	
Exported Data	[...]	
Variables	[...]	
Method	Kohonen S... ∨	
Internal Standardization	Batch SOM	
⊟ Segment	Kohonen SOM	
┊ Exported Variables	Kohonen VQ	
┊ Segment Role	Segment	
┊ Row	4	
┊ Column	2	
⊟ Seed Options		
┊ Initial Method	Default	
┊ Radius	0.0	
⊟ Batch SOM Training		
┊ Defaults	Yes	
┊ Local-Linear Smoothing	Yes	
┊ Nadaraya-Watson Smoothing	Yes	
⊟ Local-Linear Options		
┊ Convergence Criterion	1.0E-4	
┊ Max Iterations	10	

SOM/Kohonen:

In the SOM/Kohonen method, you will generate a topological map and need to determine the desired size. Generally, the larger the number of inputs, the larger the map size should be as well. In the context of CRM, however, generating 20 or more segments may not be practical or useful so this will take a bit of trial and error. Typically, the larger the map the longer it will take to train. Therefore, if you double the number of rows and columns in your map, you should double the size property in the Neighborhood Options. Choosing the SOM map size and the final neighborhood size will also take some trial and error. If the initial method in the seed options is set to principle components or outlier, etc., then you should start with a low learning rate such as 0.5. However, if you select a seed method, which is random, then the learning should be set much higher, such as 0.9. The largest learning rate is 1.0. Consult the SAS Enterprise Miner Node Reference Help for more details.

Kohonen VQ:

If you select vector quantization, then you must specify the maximum number of clusters in the Kohonen VQ property setting and the learning rate using the learning rate property. The learning rate and initial seed settings have the same properties as in the SOM method mentioned in the SOM/Kohonen method.

One of the main differences between VQ and a SOM is in the design of the network architecture. Another difference is in how the inputs are treated. In a SOM, inputs (variables of your data set) are mapped via a neural network into a two-dimensional map of size *n* rows and *m* columns. In VQ, each input is coded into a *vector* and this type of network can be *supervised,* that is trained to classify elements of a target variable, for example. In VQ, the output still produces a segment classification where each row of your data set will be classified according to the similarity of all the input vectors. In a SOM, the rows of your data set are grouped and the segment classification is the row and column location of the map produced.

Process Flow Table 1: SOM Segmentation

Step	Process Step Description	Brief Rationale
1	Create a new project called SOM Segmentation and a new process flow diagram called SOM Fuzzy Segments.	
2	Drag the CUSTOMERS data set from the Data Sources folder onto the diagram.	
3	Add a SAS Code node and connect the CUSTOMERS data set to it.	Uses the softmax macro to normalize product quantity.
4	Add SOM_DATA from SAMPSIO library to the Data Sources folder and to the process flow diagram.	Scores data containing product softmax scaling.
5	Add a Metadata node and change the levels of three variables.	Modifying levels of three variables.
6	Add an SOM/Kohonen node and connect the Metadata node to it.	Uses only the variables in Figure 11.7.
7	Modify the SOM/Kohonen property sheet to reflect the output in Figure 11.8.	Shows specific settings for SOM/Kohonen.
8	Add a Segment Profile node and keep all default settings.	Shows the additional profiling of SOM segments.
9	Copy the SOM/Kohonen node and paste it onto the diagram.	Modifies SOM settings to be VQ.
10	Add a Segment Profile node and attach the SOM/Kohonen (VQ) to it.	Shows the profile for the VQ analysis.
11	Add CUSTOMER_SCORE data to the Data Sources folder and onto diagram.	Shows data for comparing product vs. customer segmentations.
12	Add a Score node and connect both SOM nodes to the Score node.	Scores both models onto a data set.
13	Modify the SAS Code to reflect the statements shown in Figure 11.13.	Shows the frequency crosstabulations of two segments.

Step 1: So, now let's construct a couple of SOM/Kohonen segmentations. Create a new project in SAS Enterprise Miner and call it SOM Segmentation. Then you can add data sources to your project and add the data set CUSTOMERS from the SAMPSIO SAS library. Create a new process flow diagram and label it SOM Fuzzy Segments. I call SOM fuzzy segments not because the customer classification is a probability of a segment, but because we are employing a neural network where the inputs are being mapped to a single two-dimensional classification scheme; this mapping is unsupervised and not descriptive in nature. In Chapter 10, "Product Affinities and Clustering of Product Affinities," we attempted to perform some product affinity clustering and scoring using the softmax transformation of product quantities. We will revisit this again in this example.

Step 2: Drag the CUSTOMERS data source onto your flow diagram; **Step 3:** Add a SAS Code node and attach the CUSTOMERS data source to it. Open the SAS Code node and we will again add the softmax.sas macro as before and the additional SAS code as shown in Figure 11.6. The entire SAS statements are listed in Appendix 3, and the code that is included under Chapter 11 on the data CD is called softmax scoring.sas. The data step is to ensure that all product quantities are either zero or a numerical quantity as these numeric inputs will be fed into the macro computations on each product quantity. For this example, we'll use the products only and none of the product options. You could easily write another macro to loop through the variable names; however, since there are only about 17 of them, it is just as simple to cut and paste repetitive lines and change the variable name. The last softmax call then writes out the data set to the SAMPSIO library so it can be added into the data sources folder of the project. I called this output data set SOM_DATA. Now run the SAS Code node to generate the softmax scoring of the product quantities and the output data set.

Step 4: After running the SAS Code node, you should be able to add the SOM_DATA data set to your data sources folder of the project and drag it onto the SOM Fuzzy Segments flow diagram. **Step 5:** Now, add a Metadata node and connect the SOM_DATA to it. I changed the levels of the PURCHFST, PURCHLST, and YRS_PURCHASE to the levels shown in Figure 11.5.

Figure 11.5 Modified Levels of PURCHFST, PURCHLST, and YRS_PURCHASE

Name	New Role	New Level	New Report	New Order	Hide	Role
yrs_purchase	Default	Ordinal	Default	Default	No	Input
PURCHLST	Default	Interval	Default	Default	No	Input
PURCHFST	Default	Interval	Default	Default	No	Input
Prod_O	Default	Default	Default	Default	No	Input
Prod_F	Default	Default	Default	Default	No	Input
RFM	Default	Default	Default	Default	No	Input

Figure 11.6 Softmax Scoring of the Customer Data Set

```
%include 'c:\temp\softmax.sas';

data work.test; set &em_import_data;
  if prod_a=. then prod_a=0;
  if prod_b=. then prod_b=0;
  if prod_c=. then prod_c=0;
  if prod_d=. then prod_d=0;
  if prod_e=. then prod_e=0;
  if prod_f=. then prod_f=0;
  if prod_g=. then prod_g=0;
  if prod_h=. then prod_h=0;
  if prod_i=. then prod_i=0;
  if prod_j=. then prod_j=0;
  if prod_k=. then prod_k=0;
  if prod_l=. then prod_l=0;
  if prod_m=. then prod_m=0;
  if prod_n=. then prod_n=0;
  if prod_o=. then prod_o=0;
  if prod_p=. then prod_p=0;
  if prod_q=. then prod_q=0;
run;

%softmax(dsin=work.test,dsout=work.temp,var=prod_a,log=S);
%softmax(dsin=work.temp,dsout=work.temp,var=prod_b,log=S);
%softmax(dsin=work.temp,dsout=work.temp,var=prod_c,log=S);
%softmax(dsin=work.temp,dsout=work.temp,var=prod_d,log=S);
```

```
%softmax(dsin=work.temp,dsout=work.temp,var=prod_e,log=S);
%softmax(dsin=work.temp,dsout=work.temp,var=prod_f,log=S);
%softmax(dsin=work.temp,dsout=work.temp,var=prod_g,log=S);
%softmax(dsin=work.temp,dsout=work.temp,var=prod_h,log=S);
%softmax(dsin=work.temp,dsout=work.temp,var=prod_i,log=S);
%softmax(dsin=work.temp,dsout=work.temp,var=prod_j,log=S);
%softmax(dsin=work.temp,dsout=work.temp,var=prod_k,log=S);
%softmax(dsin=work.temp,dsout=work.temp,var=prod_l,log=S);
%softmax(dsin=work.temp,dsout=work.temp,var=prod_m,log=S);
%softmax(dsin=work.temp,dsout=work.temp,var=prod_n,log=S);
%softmax(dsin=work.temp,dsout=work.temp,var=prod_o,log=S);
%softmax(dsin=work.temp,dsout=work.temp,var=prod_p,log=S);
%softmax(dsin=work.temp,dsout=sampsio.som_data,var=prod_q,log=S);
```

Step 6: Now you can add a SOM/Kohonen node to the process flow diagram and connect the Metadata node to it. SOMs tend to work best with numeric data; however, in the Reference Help for Enterprise Miner 5.2, the coding for how the SOM interprets categorical data is given and is similar to how the Clustering node interprets categorical variables. If you desire to re-code them, you can use the SAS Code node or use the Replacement node to re-group levels or re-code some levels into others.

In our first SOM, the only variables we will use are the ones listed in Figure 11.7.

Figure 11.7 SOM Variables Used in Analysis

Name	Use ▽	Report	Role	Level	Type	Or...	Label	Format
PURCHLST	Yes	No	Input	Interval	N		Last Yr of Purchase	
channel	Yes	No	Input	Nominal	N		Purchase Sales Channel	
SEG	Yes	No	Input	Nominal	C		Industry Segment Code	
PURCHFST	Yes	No	Input	Interval	N		Year of 1st Purchase	
sm_prod_n	Yes	No	Input	Interval	N			
sm_prod_j	Yes	No	Input	Interval	N			
sm_prod_b	Yes	No	Input	Interval	N			
sm_prod_d	Yes	No	Input	Interval	N			
sm_prod_h	Yes	No	Input	Interval	N			
sm_prod_q	Yes	No	Input	Interval	N			
sm_prod_l	Yes	No	Input	Interval	N			
sm_prod_m	Yes	No	Input	Interval	N			
sm_prod_p	Yes	No	Input	Interval	N			
cust_id	Yes	No	ID	Nominal	C		Customer ID No.	
yrs_purchase	Yes	No	Input	Ordinal	N		No of Yrs Purchase	
sm_prod_g	Yes	No	Input	Interval	N			
us_region	Yes	No	Input	Nominal	C		US Region Location of Business	
sm_prod_i	Yes	No	Input	Interval	N			
sm_prod_c	Yes	No	Input	Interval	N			
sm_prod_k	Yes	No	Input	Interval	N			
sm_prod_a	Yes	No	Input	Interval	N			
sm_prod_o	Yes	No	Input	Interval	N			
public_sector	Yes	No	Input	Binary	N		0-No, 1=Yes	
sm_prod_e	Yes	No	Input	Interval	N			
sm_prod_f	Yes	No	Input	Interval	N			
tot_revenue	No	No	Input	Interval	N		Revenue for All Years	

The CUST_ID variable is needed to identify each customer individually (required as in Clustering), and I added a couple of demographic variables (US_REGION, YRS_PURCHASE, SEG for industry, PURCHFST, PURCHLST, and CHANNEL). Now, for the settings of the SOM/Kohonen node, there are many options on the property sheet, and fortunately for you we will not go through all of them. The options are documented, however, in the Reference Help for

Enterprise Miner 5.2. The primary option in the SOM/Kohonen node is the Method option. As discussed earlier, it has three possible selections. In this first example, we will select the Kohonen SOM in the Method property. This will produce a map, and we also will need to select the size of our map. **Step 7:** Figure 11.8 shows the SOM/Kohonen property sheet settings, with arrows indicating where I've changed the default settings. After these changes are made to the property sheet and the edit variables of the SOM/Kohonen node, you can run the node and view the results.

Figure 11.8 SOM/Kohonen Property Sheet Settings (1st Pass), Changes Marked with Arrows

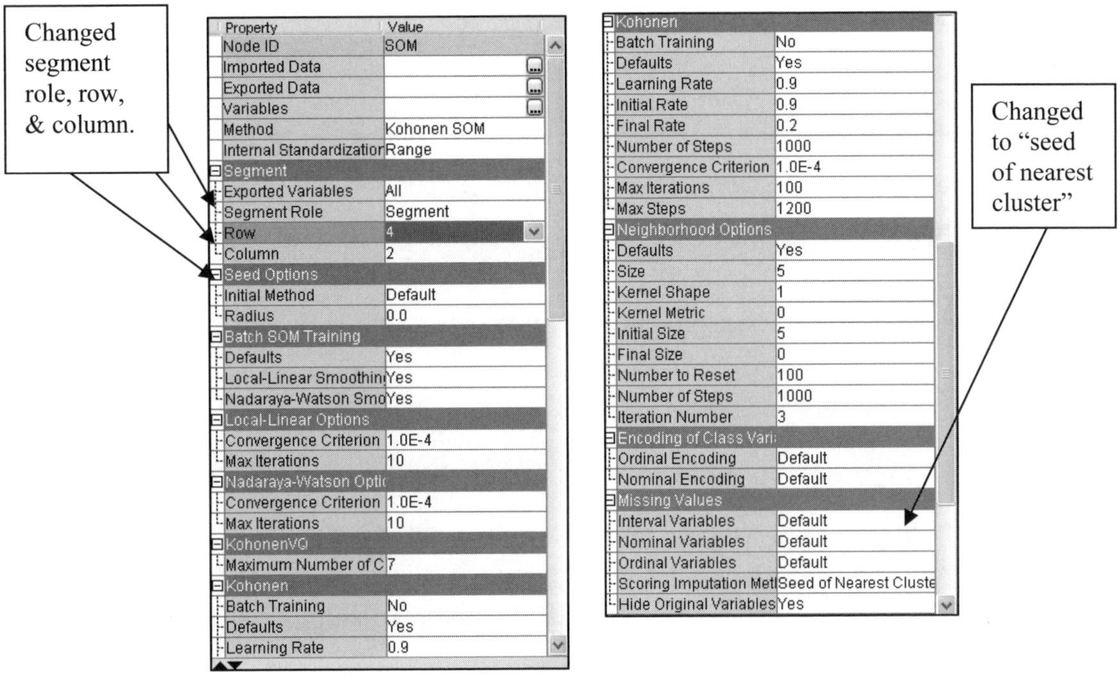

After you have run the node, the results view of the SOM/Kohonen node varies depending on the method selected. In SOM mode, a two-dimensional map is given with color intensity shading for the variable selected in a drop-down box. **Step 8:** For more in-depth profiling, however, attach a Segment Profile node after the SOM/Kohonen node and keep all defaults at this point. Run the Segment Profile node and you should now have a fairly full profile capability between the SOM map and the Segment Profile node's output. In the Segment Profile node, the Output window shows the decision tree results of importance for the variable by each segment. This should give you a very good idea of which variables are dominant in each segment. Figures 11.9 and 11.10 show the partial results of the SOM map and the Segment Profile node, respectively.

Figure 11.9 SOM Results Output

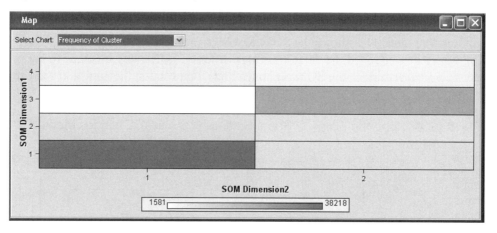

Figure 11.10 SOM Segment Profile Results Output

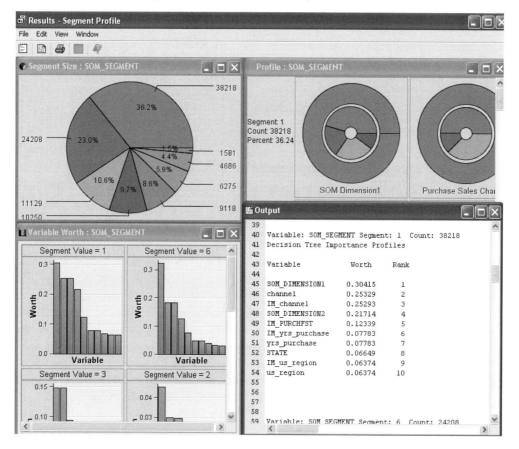

If you scroll through the Segment Profile node's Output window, each set of variables by segment should provide you with the *worth* statistic and *rank* for each variable. Notice that some variables include IM_ in front of them; these are imputed variables determined from the seed of the nearest cluster setting in the missing value property. If we selected the None option in the Scoring Imputation property, then any customer record with a missing value would be omitted in the SOM computations; however, that record would be scored with an estimate, which is not optimal. With the settings we selected, the missing values were imputed first so that computations can take place. Otherwise, the scoring would most likely not be as optimal. You could impute the values with SAS PROC MI if you desired an even more precise imputed estimate as we did earlier in Chapter 9, Clustering and the Issue of Missing Data." Each of the eight segments can be profiled and, as we've done earlier, sales or campaign planning can begin to take place with these groups.

Step 9: Now, let's try the Kohonen VQ method to see what differences take place. Copy the SOM/Kohonen node we just ran and past it onto the process flow diagram. I labeled the first SOM node SOM/Kohonen – SOM Segmentation and the second one SOM/Kohonen – Kohonen VQ Segmentation. The only changes to this one are to set the Method property to Kohonen VQ and the maximum number of clusters in the Kohonen VQ property to 8. I've kept the missing values property sheet the same as in the SOM node earlier. **Step 10:** Attach a Segment Profile node to the output of the Kohonen VQ node and run this new flow. Figure 11.11 shows the full process flow diagram to this point.

Figure 11.11 SOM/Kohonen and VQ Segmentations Flow Diagram

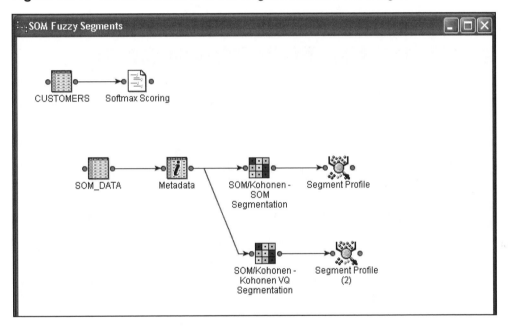

The additional exercise at the end of this chapter requires you to compare the segment profiles by product type from these two segments to see how they differ.

11.4 Comparing Clustering with SOM Segmentation

In the Chapter 5 segmentation example 5.2, we used the CUSTOMERS data set and created five customer segments using no product quantities or affinity scores but only customer demographics. We scored that segmentation on a data set called CUSTOMER_SCORE and saved it into the SAMPSIO library. **Step 11:** You should now add this data set to your data sources folder in the SOM Fuzzy Segments project. In Chapter 5, "Segmentation of Several Attributes with Clustering," we used only customer demographic variables, and in the SOM and VQ segmentation, we used product affinity scores; however, it is a useful exercise to compare how these segments appear on the same data set. There are no distance measurements written when VQ or SOM analyses are run, so we cannot compare distance-based measurements. However, we can compare the segment variables in each of the analyses.

There are a number of methods for comparing class variables with each other. The simplest is a crosstabulation of variable A with variable B, for example. In addition, a decision tree could compare all variables in the data set with variable A being a target against variable B being a target. In SAS Enterprise Miner, you have a Stat/Explore node available, which allows many multiple comparisons of numeric or classification variables. We will do the simple crosstabulation using the SAS Code node and the SAS/STAT FREQ procedure.

Step 12: In the SOM Fuzzy Segment process flow diagram, drag the CUSTOMER_SCORE data onto your diagram if you haven't done so already, and also add a Score node. Set the role of the CUSTOMER_SCORE data to Score instead of the default Raw value. Connect the CUSTOMER_SCORE data to the Score node and connect each of the SOM and VQ segmentation nodes to the Score node. **Step 13:** Place a SAS Code node after the Score node so that your process flow diagram looks like the one in Figure 11.12 and enter the SAS statements shown in Figure 11.13.

Figure 11.12 SOM Fuzzy Segment Flow Diagram for Comparing Segments

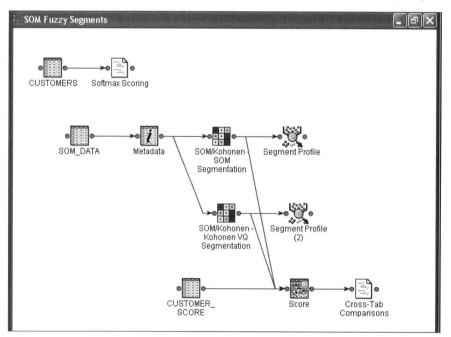

In the SAS Code node, we will run two PROC FREQ tables to compare the original customer segmentation that was scored on the CUSTOMER_SCORE data set along with the scored SOM and VQ segmentations. This, in essence, is comparing the classification scheme from the Clustering in Chapter 5 with the SOM and VQ analyses we just completed. In the SAS Code node, place the following SAS statements and run the SAS code node. I labeled the SAS Code node as Cross-Tab Comparisons.

Figure 11.13 SAS Code for Segmentation Comparisons

```
title 'Comparison of Original Segments with VQ Segments';
proc freq data=emws.score_score;

  tables _segment_ * som_segment /nocol norow nocum ;
run;
title;

title 'Comparison of Original Segments with SOM Segments';
proc freq data=emws.score_score;
  tables _segment_ * som_id /nocol norow nocum ;
run;
title;
```

After the SAS Code node has run, you should see two crosstabulation frequency distribution reports in the Results Output window. Figure 11.14 shows the output results from the preceding SAS Code node statements. What we observe in Figure 11.14 is how each customer record is classified in these segments. Notice that in the original comparison (_segment_ variable) the original cluster segment 2 is very similar to the VQ segment level 1. In addition, VQ segment number 6 is also very similar to original segment 3. The overall summary results are tabulated in Tables 11.1 and 11.2 with the original segments to VQ and SOM segments, respectively.

Table 11.1 Original Clusters vs. VQ Segments Summary

Original cluster segment	Similar to VQ segments
1	6, 2, 8
2	1, 4, 7
3	6, 1, 2
4	1, 6, 4
5	3, 1, 7

Table 11.2 Original Clusters vs. SOM Segments Summary

Original cluster segment	Similar to SOM segments
1	3:2, 4:2, 2:1
2	1:1, 2:2, 4:1
3	3:1, 1:1, 4:2
4	1:1, 3:2, 2:2
5	2:1, 1:1, 4:1

Figure 11.14 Original vs. SOM and VQ Segmentation Comparison Results

```
Comparison of Original Segments with VQ Segments

Table of _SEGMENT_ by SOM_SEGMENT

_SEGMENT_       SOM_SEGMENT(SOM Segment ID)
```

Frequency Percent	1	2	3	4	5	6	7	8	Total
1	0	3080	1175	0	97	6825	374	1527	13078
	0.00	2.92	1.11	0.00	0.09	6.47	0.35	1.45	12.40
2	16747	1	551	3935	498	0	1901	0	23633
	15.88	0.00	0.52	3.73	0.47	0.00	1.80	0.00	22.41
3	5416	4432	500	1206	216	11864	755	1956	26345
	5.14	4.20	0.47	1.14	0.20	11.25	0.72	1.85	24.98
4	10097	1786	2381	2526	374	3768	1439	747	23118
	9.57	1.69	2.26	2.40	0.35	3.57	1.36	0.71	21.92
5	5965	951	6521	1451	396	1745	1807	455	19291
	5.66	0.90	6.18	1.38	0.38	1.65	1.71	0.43	18.29
Total	38225	10250	11128	9118	1581	24202	6276	4685	105465
	36.24	9.72	10.55	8.65	1.50	22.95	5.95	4.44	100.00

```
Comparison of Original Segments with SOM Segments

Table of _SEGMENT_ by SOM_ID

_SEGMENT_       SOM_ID(SOM ID)
```

Frequency Percent	1:1	1:2	2:1	2:2	3:1	3:2	4:1	4:2	Total
1	0	3080	1175	0	97	6825	374	1527	13078
	0.00	2.92	1.11	0.00	0.09	6.47	0.35	1.45	12.40
2	16747	1	551	3935	498	0	1901	0	23633
	15.88	0.00	0.52	3.73	0.47	0.00	1.80	0.00	22.41
3	5416	4432	500	1206	216	11864	755	1956	26345
	5.14	4.20	0.47	1.14	0.20	11.25	0.72	1.85	24.98
4	10097	1786	2381	2526	374	3768	1439	747	23118
	9.57	1.69	2.26	2.40	0.35	3.57	1.36	0.71	21.92
5	5965	951	6521	1451	396	1745	1807	455	19291
	5.66	0.90	6.18	1.38	0.38	1.65	1.71	0.43	18.29
Total	38225	10250	11128	9118	1581	24202	6276	4685	105465
	36.24	9.72	10.55	8.65	1.50	22.95	5.95	4.44	100.00

In summary, the VQ and the SOM methods for classification can be used when there are many inputs and the relationships between the variables are rather complex. The deciding factor as to when to use SOM or VQ as a segmentation technique may depend somewhat on the following key attributes:

- The data contain many variables of complex relationships.

- Ordinary *k*-means clustering does not appear adequate or does not produce desired results.

- The priority is not very high to explain the nature of how variables were used in the segmentation.

The distinction of VQ and SOM in the computer science and information theory literature is not very clear. However, when faced with the issues of non-normal data and data that perhaps has complex relationships, you can use one or more of these techniques in SAS Enterprise Miner. You may have to try one or more of these algorithms to see what works best in your situation.

11.5 Customer Distinction Analysis Example

If you have performed a survey of your customer's attitudes towards their likes, dislikes, how they prefer to do business, whether they like or dislike email newsletters or white papers that help them in their business decisions, etc., then the responses from your customer surveys can be segmented for various groups of customers that have similar preferences. In addition, if your selection was done according to random sampling within each set of RFM cells (or a similar segmentation), then you will have a very valuable set of customer attributes in which to help you make better CRM business decisions.

Process Flow Table 2: SOM Segmentation

Step	Process Step Description	Brief Rationale
1	Reopen the project RFM Cell Development from Chapter 4, and add a new diagram.	
2	Add a SAS Code node and enter the SAS statements.	Creates a subset of data on specific RFM levels.
3	Add a Sampling node to the diagram and connect the Code node to it.	Stratifies sampling for 500 customers only.
4	Drag a SOM/Kohonen node and set the property sheet values.	
5	Add BUYTEST data from the SAMPSIO library to the data sources folder.	Sets Role of the data set to Score.
6	Add a Score node and attach the SOM.	
7	Add a Segment Profile node and attach the SOM node to it.	Shows the full diagram in Figure 11.18.

Step 1: Open SAS Enterprise Miner and open the project called RFM Cell Development that we formulated back in Chapter 4, "Segmentation Using a Cell-based Approach." Now, create a new diagram and call this new diagram Customer Distinction. Add a data set from the SAMPSIO library called RFM_SCORE_TEST. This is the TEST data set we scored in the RFM Cell exercise. Drag this new data set from the Data Sources folder to the process flow diagram in the Customer Distinction process flow. **Step 2:** Now attach a SAS Code node and open the code in the properties sheet. Use the SAS code shown in Figure 11.15 to exclude the following RFM values: B, C, E, F, H, I, J, and L.

Figure 11.15 SAS Code to Exclude RFM Values

```
data &EM_EXPORT_SCORE;
  set &EM_IMPORT_DATA;
    where RFM not in ('B','C','E','F','H','I','J','L');
run;
```

Then close the SAS Code node. This will only pass on to remaining nodes any records that do not have the RFM score values that were selected. **Step 3:** Next drop in a Sampling node and connect it from the SAS Code node. Highlight the Sampling node. In the property sheet, select Stratified for the sampling method, and set the number for the sample size to 500. This will be the number of survey customers. In the Variables sheet, select both the RFM and the RESPOND variables and set them to Stratify. Set the other fields to default. This flow ensures that when the random samples are performed, the RFM levels and the Response levels are sampled proportionately to their actual distributions. We need to ensure that all the RFM and Response levels are correctly sampled so that when scoring is done back in the original data set the proportions are correct. Now connect a SAS Code node and in the program window place the survey.sas code from the Chapter 11 folder of your CD. The code is listed in Appendix 3.

Step 4: Now connect a SOM/Kohonen node and the SAS Code node to the SOM node. Each geographic portion of the map corresponds to a cluster segment. In the SOM property sheet, set the Method to Kohonen SOM and the internal standardization to Range. Select Variables and choose only RFM, ID, AGE, INCOME, SEX, and OWNHOME, LOC, CLIMATE, and Q1–Q5; do not select the other variables. In the Segment portion of the property sheet, set the Kohonen SOM to 4 rows and 4 columns. Leave the other settings at their default values. Now you can run the SOM node after you close this node and save your changes. Remember what we've done here is to add a five-question questionnaire to 500 customers from randomly selected RFM and Response values. The end result should be to score the question responses with the RFM onto the remainder of the data set of 10,000 records. The SOM map should look like the one shown in Figure 11.16. You can select the chart type in the upper left menu and observe how each variable is segmented in the SOM map. The darker the shading, the higher the concentration of the variable in that portion of the map. In order to score these 500 observations with the SOM model, we'll need the full data set.

Figure 11.16 SOM Cluster Map from Attribute and Demographic Variables

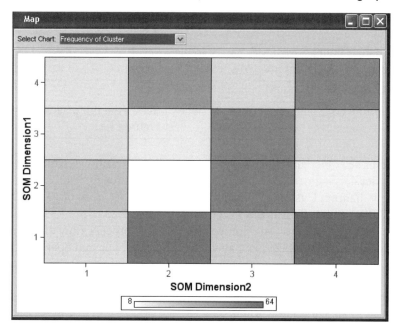

Step 5: Now, drag the BUYTEST data onto the diagram and set the role of this data to SCORE in the properties sheet. **Step 6:** Add a Score node and connect the SOM/Kohonen output to it and also the BUYTEST. This will score the SOM segment mode on all the BUYTEST records even though we had only our questions on the 500 survey sample. **Step 7:** Also attach a Segment Profile node to the output of the Score node. This will aid in giving you some profile information about the segments that the SOM/Kohonen network found. In the Segment Profile node, set the value in the Use column to No for the selected variables as shown in Figure 11.17. Set this value to Yes for all other variables.

Figure 11.17 Variable Selection for Segment Profile Node

Name	Use	Report	Role	Level
EM_CLASSTARGET	No	No	Assessment	Nominal
F_RESPOND	No	No	Classification	Binary
ORGSRC	No	No	Input	Nominal
EM_PROBABILITY	No	No	Prediction	Interval
EM_EVENTPROBABILITY	No	No	Prediction	Interval
I_RESPOND	No	No	Rejected	Unary
Distance	No	No	Rejected	Interval
EM_CLASSIFICATION	No	No	Rejected	Nominal
VALUE24	No	No	Input	Interval
survey_score	No	No	Input	Interval
R_RESPOND1	No	No	Residual	Nominal
V_RESPOND1	No	No	Prediction	Nominal
SOM_DIMENSION2	No	No	Input	Nominal
SOM_DIMENSION1	No	No	Input	Nominal
U_RESPOND	No	No	Rejected	Unary
NODE	No	No	Input	Nominal
V_RESPOND0	No	No	Prediction	Nominal
R_RESPOND0	No	No	Residual	Nominal
P_RESPOND0	No	No	Prediction	Nominal
WARN	No	No	Rejected	Unary
RESPOND	Yes	No	Input	Binary

The SOM/Kohonen node will attempt to place the survey question data along with the demographics and the RESPOND variable into a cluster map of two dimensions. We would like to see what cluster and survey components combined with any demographic variables for more descriptive information of the customers who answered the questions in certain ways. By including them in the analysis of the 500 respondents and developing a cluster model, we can then score the remainder of the database with such a model. Figure 11.18 shows the completed process flow diagram.

Figure 11.18 Customer Distinction Process Flow Diagram

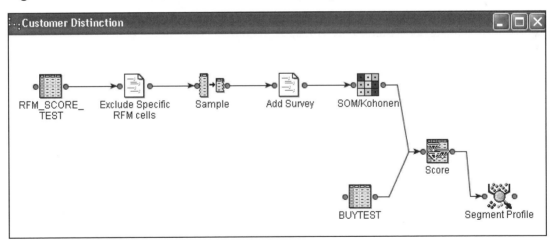

So, let us recap what we've accomplished in this data mining process flow.

- We took the test data set from our RFM diagram where we created RFM cells based on the values of several purchase pattern variables and randomly sampled 500 customers stratified by certain RFM cells for a customer survey.

- We then appended the 500-customer survey results back into the data flow and combined them with the demographic elements; we created an SOM segmentation model of survey questions and demography for the 500 customers.

- Then, with the SOM segmentation model, we scored the data set of 10,000 customers with the model where the majority didn't have customer survey responses.

- Now the completed data set contains RFM cells and a prediction of customer distinction obtained from the 500 survey customer analysis.

What can you do now with these new segments? Once each segment is again profiled as shown earlier, specific programs and campaigns can be developed for each customer segment or scored segment, and thus a methodology of CRM by customer segment can now begin starting from data and information that exists in your database. In summary, in this chapter we've gone over a number of cases that show how you can classify customers according to their RFM scores. We discussed using these scores to better understand your customer base sampling a set of customers according to their RFM values, surveying these customers for attitudinal information, and then scoring the remainder of the database as if you had run the survey on the entire database.

This methodology was used in this case for the following reasons. With 500 customer responses on the survey, there may not be enough data records to build an ordinary predictive model using a regression, decision tree, or neural network. This is typical for customer survey data in general. In light of this, I chose to run a segmentation model that incorporated the responses along with their demography and then scored those segments on the complete data set. This technique allows the segments to be profiled that have pockets of customers who answered the questions. The hope is to have these segments differ from one another in responses and demography so that each segment can be marketed to according to responses or customers who might be likely to respond similarly (e.g., customers in the same segment).

11.6 Additional Exercise

In Section 11.3, process flows exist for both the SOM segmentation and a VQ segmentation. Write a SAS code that does not compare the softmax-scaled product affinities but does compare the mean values of products by each segment for these two segmentation schemes. Comment on the similarities and/or differences in the segments.

11.7 References

Kohonen, T. 1981. "Automatic formation of topological maps of patterns in a self-organizing system." In Oja, E., and O. Simula, eds. *Proceedings of the Second Scandinavian Conference on Image Analysis*. Helsinki, Finland: Suomen Hahmontunnistustutkimuksen Seura r. y. 214–220.

Kohonen, T. 1988. "Learning Vector Quantization." *Neural Networks*. 1 (Suppl. No. 1):303.

SAS Enterprise Miner, Release 5.2. Cary, NC: SAS Institute Inc.

Segmentation of Textual Data

12.1 Background of Textual Data in the Context of CRM

It has been said that about 80% of a corporation's data is contained in textual form of documents, e-mails, and other unstructured or semi-structured text (Sullivan 2001; Hearst 1999). This being the case, it is no wonder that the text mining market has boomed as vendors compete for this lucrative analytics market. Think about how much data your company has in textual form: documents on your internal Web site, e-mail messages, documents in PDF, analysis or health record reports, medical diagnoses, financial reports, customer notes and feedback forms, problem reports, and so on. Just searching on Google for "text mining, text analytics" will produce a query of about 22 million Web page references. Obviously, even the best speed reader cannot browse this sort of document volume and have any serious mental recall of the topics that were read or even classified.

The need to expand business intelligence and analytical intelligence using textual information is greater now than ever before. There are several reasons why this is the case. First, the set of tools at your disposal for analyzing textual data is now commercially available. SAS Text Miner allows you to analyze textual data along with structured data in a common data mining tool set in SAS Enterprise Miner. SAS Text Miner is an add-on application to SAS Enterprise Miner and if you don't have SAS Text Miner on your system, that is OK, as you can still follow the basic ideas in this chapter and understand how to apply text mining. It is advantageous if you do have it installed, however. Second, because of the Internet, there is so much data that can be researched and utilized; however, you need to be careful because some Web sites have a clause indicating that downloading without their permission is unlawful. Third, with the amount of textual data that exists in most organizations, the means of dealing with this volume of textual information are no longer sufficient to meet the needs of business decision makers. In other words, if you're not dealing with this data in an intelligent fashion, then it is likely that you are falling behind the business intelligence curve and are probably losing valuable information you might not even know about.

With the recent addition of buzzwords like *text mining*, *text analytics*, *search analytics*, etc., some definitions about what is text mining might be in good order. There are several related disciplines that have been in the literature and those in particular include the following:

- information retrieval (IR)
- computational linguistics
- document classification
- natural language processing (NLP)

While these disciplines can and do deal with textual data, they are all not intended for the same purpose. Since text mining is still a relatively new discipline, it is sometimes difficult to obtain a generally agreed-upon definition. A rather broad definition would try to identify text mining with any process or operation related to the gathering and analyzing text from external sources for business intelligence purposes (Sullivan 2001, pp. 4, 324–326). Another definition takes on a mining metaphor where text mining is the discovery of previously unknown knowledge in text. Sullivan (2001, p. 326) defines text mining a little more formally: "Text mining is the process of compiling, organizing, and analyzing large document collections to support the delivery of targeted types of information to analysts and decision makers and to discover relationships between related facts that span wide domains of inquiry." While this is a bit long winded, it does make clear that text mining can be just trying to find relationships in related facts or part of something larger like building a predictive model using text and structured data.

In CRM applications, text mining can typically be classified into one of two basic goals: prediction of something using textual data or searching for specific information in a large volume of text to uncover desired information. The latter is typically performed by clustering documents into similar topics or themes based on their content and then profiling the clusters as we did earlier using structured data. Sometimes a goal might be wanting to understand what my most valuable customer segment is speaking about to our call center representatives and asking whether these topics are different from those of other less valuable customer segments. Based on the findings of such analyses, specific offers, communications, etc., can be devised for each group according their derived set of needs. At other times, the textual data might just as well be inputs into a model to aid in the prediction of some categorical or numeric variable of interest. These are two very different applications and typically require different renditions of how the textual data is post-processed after the parsing stage. We will discuss some of these renditions next.

12.2 Notes on Text Mining versus Natural Language Processing

Textual documents are often represented efficiently by using what is called the *vector space model*. From our discussions in Chapter 3, we learned that in order to cluster data points, we needed to measure their distance from one another. The *vector space model* depicts a document in vector space by first parsing the text into parts of speech, such as verbs, nouns, noun groups, conjunctions, adjectives, prepositions, and the like. Then, once the text is parsed in this fashion, these terms are counted and recorded for each document. This forms a document-term matrix similar to that shown in Table 12.1. Documents can then be grouped or clustered according to similar sets of terms and therefore form document themes.

Table 12.1 Document-Term Matrix in an Example Seven-Document Set

Document	Term 1	Term 2	Term 3	Term 4	Term 5	Term 6
Doc 1	2	0	1	1	3	2
Doc 2	0	1	2	0	1	0
Doc 3	0	0	3	2	0	4
Doc 4	1	2	0	0	2	0
Doc 5	1	3	0	5	3	1
Doc 6	4	0	0	0	1	0
Doc 7	2	0	3	0	0	3

Let's take an example taken from Sullivan (2001, p. 324). A text mining researcher found that human deficiencies of magnesium might be related to migraines when extracting attributes from a large set of medical literature on migraines and nutrition. Subsets of extracted snippets of information from Hearst (1999) are as follows:

- Stress is sometimes associated with migraines.
- Stress can lead to a loss of magnesium.
- Calcium channel blockers prevent some migraines.
- Magnesium is a natural calcium channel blocker.
- A form of depression called spreading cortical depression (SCD) is interlaced in some migraines.
- High levels of magnesium tend to inhibit SCD.
- Migraine patients have high platelet aggregability.
- Magnesium can suppress platelet aggregability.

These series of facts were discovered by using text mining techniques; however, the link of magnesium to migraines was also supported from research experimentation as well.

If these facts could be represented in a few simple queries of keywords such as MAGNESIUM AND CALCIUM AND MIGRANE, then we would want a set of documents to be represented in a similar fashion.

DOCUMENT 1:   MAGNESIUM AND ZINC AND HYPERTENSION

DOCUMENT 2:   MIGRAINE AND SLEEP DEPRIVATION

DOCUMENT 3:   HYPERTENSION AND SODIUM AND CALCIUM

So now, instead of searching for magnesium, calcium, and migraine sequentially, we should be able to do it in a single operation. In order to accomplish this query then, three sets of facts need to be represented as an object or perhaps a numeric quantity. If we assume we are interested in documents about migraines and magnesium, then we have four possible combinations that describe documents:

- documents **about** *migraines* and **about** *magnesium*

- documents **about** *migraines* and **not about** *magnesium*

- documents **not about** *migraines* but **about** *magnesium*

- documents **not about** *migraines* and **not about** *magnesium*

We can chart these sets of two-term combinations into points on a two-dimensional axis. Now, as we add a little complexity, if we measure the relative frequency of the terms *migraines* and *magnesium,* and if *migraine* has a higher frequency, then the metric used should correspond to reflect that weight. Referring back to our term-document matrix in Table 12.1, if each term now represents the weights of the relative frequencies, then mathematical vectors can now represent this form. Extending this example further, if the terms in Table 12.1 are enumerated in Table 12.2, then term vectors that indicate the relative weight of significant terms in each document might be represented as in Table 12.3.

Table 12.2 Sample Index of Documents by Terms for Migraine and Magnesium Example

Document	Term 1: Migraine	Term 2: Magnesium	Term 3: Calcium	Term 4: Platelet	Term 5: Hypertension	Term 6: Steroid
Doc 1	2	0	1	1	3	2
Doc 2	0	1	2	0	1	0
Doc 3	0	0	3	2	0	4
Doc 4	1	2	0	0	2	0
Doc 5	1	3	0	5	3	1
Doc 6	4	0	0	0	1	0
Doc 7	2	0	3	0	0	3

Table 12.3 Example Term Vectors That Indicate Relative Term Weights in Documents in Table 12.2

Document Topic	Term Vector
Migraine, magnesium, platelet	(0.8, 0.6, 0.3, 2.1, 0, 0, 0.1, 0 , 0,....)
Platelets, hypertension	(0, 0, 0, 0.93, 0.72, 0, 0, 0, 0, 0, 0 ...)
Calcium, hypertension	(0, 0, 0.85, 0.28, 0, 0, 0, 0, 0.3, 0, ...)
Migraine, steroid, calcium	(0.78, 0, 0.52, 0, 0, 0.84, 0, 0, 0, ...)

The actual data in both Tables 12.2 and 12.3 are only example representations of how data can be transformed from text parsing of important terms, to relative frequencies, to vectors, and so on. The measure of similarity we discussed in Chapter 3, "Distance: The Basic Measures of Similarity and Association," can be applied as we calculate the angle of the vectors as in Equations 3.2 and 3.3. Documents that have similar themes can be clustered based on the distances of term vectors. These illustrations show you one of the techniques used to analyze textual data.

Fortunately, in SAS Text Miner, we don't have to do these computations by hand! In SAS Text Miner, the words are parsed and a term-by-document matrix is automatically computed. A custom synonym list can be used to aid this process because technical terms not normally recognized as synonymous, like gigabyte, disk drive, and disk storage might be considered as synonyms in some textual applications. SAS manipulates the term-by-document matrix and performs some mathematical computations using a statistical technique called Singular Value Decomposition (SVD) (SAS Institute Inc. 2005).

12.3 Simple Text Mining Example

In CRM applications, text items such as sales documents, notes, and comments from customers, surveys, blogs, news clips, and the like are all potential sources of information that could be mined via text mining. E-mail messages and Web pages (within certain legal constraints) are also potential candidates. We will begin our first example with a simple textual example of 600 news stories, which range from a paragraph to a couple of pages in length. This example is provided in SAS Text Miner documentation as well. The data set is found in the SAMPSIO library and is called NEWS. There are four variables to this data set: the first is TEXT, which contains the actual textual data of the news story; GRAPHICS, HOCKEY, and MEDICAL are all binary variables, which are numerical classifications of the documents belonging to one of the three categories respectively. If HOCKEY is a 1, then that document belongs to the topical category of Hockey, and 0 otherwise. The same classification is likewise for Graphics and Medical. So, in the NEWS data set, each document was obtained and manually classified by the three general topics of Hockey, Medical, and Graphics.

Process Flow Table 1: Text Segmentation—News Stories

Step	Process Step Description	Brief Rationale
1	Create a new Project called Text Segmentation and a new diagram called News Stories.	
2	Add a data set from the %SAMPSIO library called NEWS.	Sets the HOCKEY variable to target.
3	Add a Data Partition node using the default values (60%, 20%, 20%).	Splits into training, validation, and test data.
4	Add a Text Mining node.	Mines the news articles.
5	Add a Memory Based Reasoning (MBR) node.	Classifies Hockey from news stories.
6	Add a SAS Code node following the MBR node.	Computes and plots an ROC chart.

Step 1: So, let's create a simple textual analysis on this data set. Create a new project called Text Segmentation and add a new data set to the Data Sources folder; this is the NEWS data set and can be found in the SAMPSIO SAS library. Create a new diagram called News Stories. **Step 2:** Once you have added it to the Data Sources folder, drag it onto a new process flow diagram called News Stories. Open the Variables property sheet, set the HOCKEY variable's role to TARGET, and reject the variables GRAPHICS and MEDICAL for now. Ensure that the variable called TEXT is set to a role of TEXT. Now add the following nodes and modify only the attributes selected:

Step 3: Add a Data Partition node.

- Connect the NEWS data source to the Data Partition node. Set the partitioning Method property to Simple Random.
- In the properties panel of this node, set the data set percentages of the training, validation, and test sets to 60%, 20%, and 20% respectively.

Step 4: Add a Text Mining node.

- Connect the Data Partition node to the Text Mining node.
- Be sure that the Stem Terms, Different Parts of Speech, and Noun Group properties are all set to Yes.
- Use the default settings on everything else; however, set the Max SVD dimensions to 50 instead of to the default value of 100.

Step 5: Add a Memory Based Reasoning (MBR) node:

- Connect the Text Mining node to the MBR node.
- Use all default settings for this node.

You can now run the MBR node. In the Results window of the MBR node, you should be able to pull down the View menu and select Assessment, and then Classification Chart: Hockey. Figure 12.1 shows the resultant chart. It shows how well the MBR model classified the binary variable HOCKEY.

Figure 12.1 Classification Chart for the Binary HOCKEY Variable

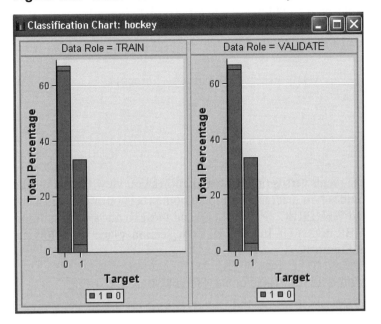

Let's review what we have done so far. The 600 news stories have been partitioned randomly into training, validation, and test sets. The Text Mining node has parsed the text of these stories, and we have built an MBR model to predict the predetermined classification of Hockey, which is the correct classification of news articles that are about Hockey. The chart in Figure 12.1 shows the percentages of correct classification for the training and validation sets. The chart for validation shows that the model correctly classifies the Hockey variable with 1 about 65% of the time.

In the information retrieval industry, precision and recall are important text document metrics. Precision and recall are measures that describe how effective a binary text classifier predicts documents that are relevant to a particular category. Recall measures how well the classifier can find relevant documents and properly assign them to their correct category. Precision and recall can be computed from a crosstabulation table (also called a contingency table) as shown in Table 12.4 (SAS Institute Inc. 2005).

Table 12.4 Contingency Table of Actual vs. Predicted Classifications

	Predicted Value 1	**Predicted Value 0**
Actual 1	α	β
Actual 0	γ	δ

If your interest in the target variable is 1, then the value in cell A of Table 12.3 is the number of correct documents predicted that actually belong to that group where $\alpha + \gamma$ are the *total* documents belonging to that group. Precision and recall can be computed in the following formulas with respect to Table 12.3.

$$\text{Precision} = \frac{\alpha}{\alpha + \beta}$$

$$\text{Recall} = \frac{\alpha}{\alpha + \gamma}$$

An ROC chart is a graph of the recall versus precision and allows you view the trade-off of precision and recall. SAS provides two macros to compute both precision and recall. These macros are %PRCRCROC and %PREREC. **Step 6:** Let's add these to our analysis; drag a SAS Code node and connect the MBR node to it. In the SAS Code section, place the following statements as shown below in Figure 12.2.

Figure 12.2 SAS Code Statements for Precision vs. Recall Chart and Table

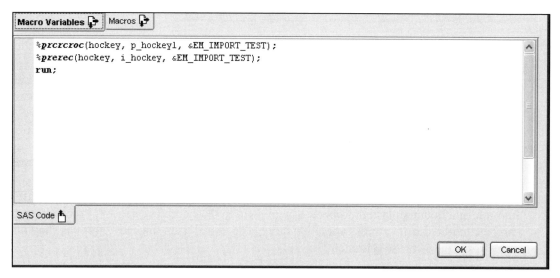

Close the SAS code section and run the SAS Code node. Figure 12.3 shows the SAS Code output from the SAS Code node and the menu to obtain the Recall vs. Precision graph. The Recall vs. Precision graph is shown in Figure 12.4 and can be selected from the SAS Graphs menu option.

Figure 12.3 SAS Code Node Output with Menu for ROC Chart

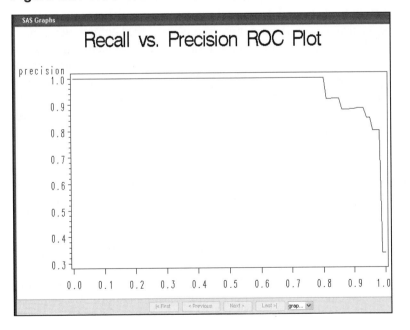

```
Results - ROC Chart

File  Edit  View  Window

      Properties        ►
      SAS Results       ►    Log
      Scoring           ►    Output
      Table                  Flow Code
      Graph Wizard           SAS Graphs

                               Output
                            22  Recall vs. Precision ROC Plot
                            23
                            24  The FREQ Procedure
                            25
                            26  Table of hockey by I_hockey
                            27
                            28  hockey       I_hockey(Into: hockey)
                            29
                            30  Frequency|
                            31  Percent  |
                            32  Row Pct  |
                            33  Col Pct  |0        |1       |  Total
                            34  ---------+--------+--------+
                            35        0  |    78  |     3  |     81
                            36           |  63.93 |   2.46 |  66.39
                            37           |  96.30 |   3.70 |
                            38           |  91.76 |   8.11 |
                            39  ---------+--------+--------+
                            40        1  |     7  |    34  |     41
                            41           |   5.74 |  27.87 |  33.61
                            42           |  17.07 |  82.93 |
                            43           |   8.24 |  91.89 |
                            44  ---------+--------+--------+
                            45  Total         85       37      122
                            46               69.67    30.33   100.00
                            47
                            48
                            49
                            50  Recall vs. Precision ROC Plot
                            51
                            52  Obs   clasrate   misclass   precision   recall   breakeven
                            53
                            54   1   91.8033    8.19672    0.91892    0.82927   0.87409
                            55
```

Figure 12.4 ROC Chart from SAS Code Node Macro Statements

From the analysis that we done so far, the following statements can be made:

- 78 articles out of the 122 (or 63.93%) are correctly classified as 0, meaning not about Hockey.

- 34 of the total of 37 articles (91.89%) are correctly classified as 1, meaning they are about Hockey where the I_Hockey variable is equal to 1 (the predicted Hockey variable).

- From the combined set of articles that belong to the Hockey category (41), 10 are misclassified (or 24.4%).

- The break-even point of the ROC chart is about 0.92, which is the average of precision and recall.

Now, what can we say about the business context of these news stories? Predicting which news articles are about hockey, medicine, or graphics is not very impressive if it is not being applied to solving some sort of business problem or issue. Predicting a correct class of document is useful, if applied to the proper business setting. If the text documents were notes from a call center instead of news stories, and the classification variable to be predicted was an attitudinal segment of customers, then this predictive model application could be applied to the remainder of the database of customers where the attitudinal segmentation does not exist.

12.4 Text Document Clustering

Sometimes, the business question or issue at hand is related to information discovery. In business competitive intelligence, one of the main goals is to uncover information about what the competition is doing. For example, in the pharmaceutical industry, searching for the kind of patents the competitive company is developing and the particular claims those patents have is useful information when forming a strategy for new drug development. Text mining in competitive intelligence can also be used for finding terrorist activities among incoming Internet documents. There are many such examples of textual information available to business, industry, and academia. One of the ways in which information is *discovered* is to cluster documents into potential themes. The application discussed earlier where a text mining researcher found some relating facts about migraines and nutrient deficiencies might have used document clustering to find documents that were similar to each other in their content themes. We discussed briefly some of the mechanics of how documents can be clustered by measuring distances of documents using the relative weights of terms in each set of documents.

Cluster profiling on document groups can be time consuming and is more difficult in general than the cluster profiling we discussed in earlier chapters due to the nature of textual data. In SAS Text Miner, there are a couple of techniques used to aid in the profiling of segments from textual documents that have been clustered. First, there is a keyword descriptive term summary. In the Text Mining node, the Cluster property sheet contains a Descriptive Term item, in which you can set the number of terms (keywords or noun-group phrases) to be used; the default is set to five. Secondly, several graphical displays in the Text Miner Node Results window show a number of charts that can aid in the understanding of the mining results. From our earlier example of news stories, the Results window is shown in Figure 12.5 and displays the frequency in number of terms by their grammatical role, by term weight, etc. Thirdly, instead of opening the Text Mining node results, the property called Interactive opens an Interactive Results window. This window shows the term-document matrix with terms classified according to their part of speech role and the view of the textual document data as well. Columns in the term-document matrix table can be sorted by clicking the column heading (sorts in ascending or descending sequence). If you select a term of interest, such as player in the previous example, then you can right-click to pull up a menu of several options. Terms can be removed from the analysis; terms that are similar can be viewed or considered as synonymous with other terms selected. In addition, concepts linked to the highlighted term can be viewed with a concept link graph. This concept link graph is dynamic and not static! In the news stories analysis, highlight the Text Mining node and open the Interactive window. Sort the terms and highlight the term Player. Now, choose the Select View Concept Links. A window opens showing concepts, which are linked to the term in the center. This is a highly interactive window where you expand individual links further, or move the entire link graph geographically. In this example, I selected the link called Season, and I moved the concept link so that the link graph views as shown in Figure 12.6.

Figure 12.5 Text Mining Results Window from News Stories

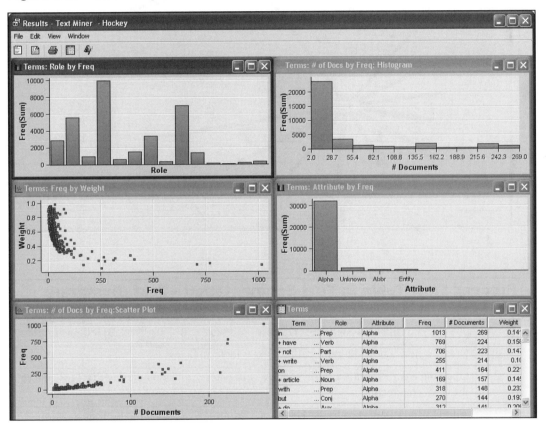

Using the concept link graphic, you can view the terms linked to others visually and aid the miner in textual information discovery. The width of the drawn link shows the relative strength of the link association. Descriptive terms that are given for each cluster can profiled in this fashion for concepts that are linked together.

Figure 12.6 Player–Season Concept Link Graph in News Stories Text Mining Example

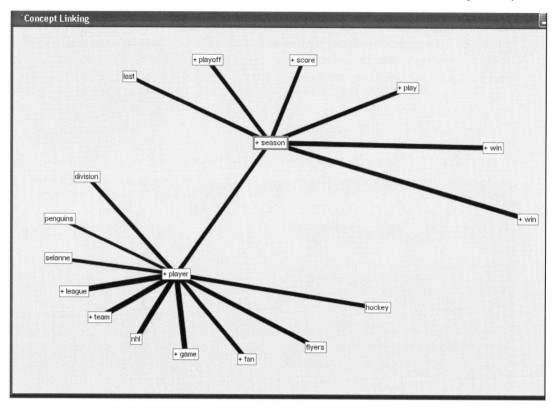

Process Flow Table 2: Text Segmentation—Text Clustering

Step	Process Step Description	Brief Rationale
1	Create a new a new diagram called Text Clustering.	
2	Open the SAS Program Editor window and enter the SAS statements.	Runs the TMFILTER macro to Web crawl.
3	Add DMREVIEW data to the Data Sources folder after running TMFILTER.	Adds the pointer location of the Web pages on your hard drive.
4	Add a Text Mining node and set up properties for parsing and clustering.	Attempts to cluster Web documents.
5	View the Text Interactive Results window.	Shows what text clusters were found.
6	Revise the Text Mining node, removing parts of speech and re-clustering.	Gets rid of low information words.
7	Profile resulting re-clusters from Text Miner node using the interactive window.	

Step 1: Let us now attempt a business application like crawling the Web for articles on the DMReview.com Web site. This Web site contains articles about data mining, data quality and cleansing, CRM, and the like. To accomplish this function, SAS has provided a simple tool called %TMFILTER macro. This is provided with SAS Text Miner and allows several text parsing and Web site crawling capabilities. There is some online SAS Enterprise Miner documentation under the Text Miner Node Reference. To start our process of data collection, you will need an Internet connection. **Step 2:** To start the %TMFILTER macro, open the SAS Code in the upper right corner of the SAS Enterprise Miner window as shown in Figure 12.7.

Figure 12.7 Location of SAS Program Editor Node

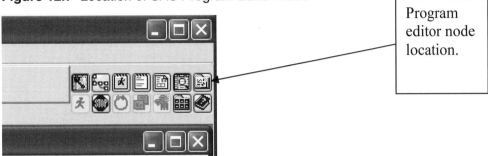

Program editor node location.

TMFILTER Macro Call for Web Crawling

The following macro call locates all allowable Web pages in the address **http://dmreview.com** to a depth of three and places them in the directory folder labeled **c:\temp\business**:

```
%tmfilter(url=http://dmreview.com,
depth=3,dir=c:\temp\business,destdir=
c:\temp\destination,norestrict=,dataset=sampsio.dmreview);
```

The filtered Web pages then are placed in the folder **c:\temp\destination**. The norestrict option set to blank implies that only documents within the specified Web domain will be processed. If this option is set to anything other than blank, then reference links within the Web domain that point to other Web pages outside of that domain will be included in the extraction. There is a rather large difference in time required for extraction if the number of reference links to outside Web pages increases. At the time of this book's writing, the TMFILTER macro extracts about 3200 Web documents from the DMReview.com Web site. This should take about 20 minutes or so depending on your Internet speed and computing environment.

You now have the pages extracted and filtered into directories on a disk system and the data set DMREVIEW in the SAMPSIO library will contain reference pointers to these pages on the disk in a variable called FILTERED. Figure 12.8 shows the first 100 rows of what the data set looks like. Since this data set does not contain the actual text documents, SAS Text Miner will know from the variables in this data set to look for them in the folder specified.

Figure 12.8 Sample of the DMREVIEW Data Set from the TMFILTER Macro Extraction

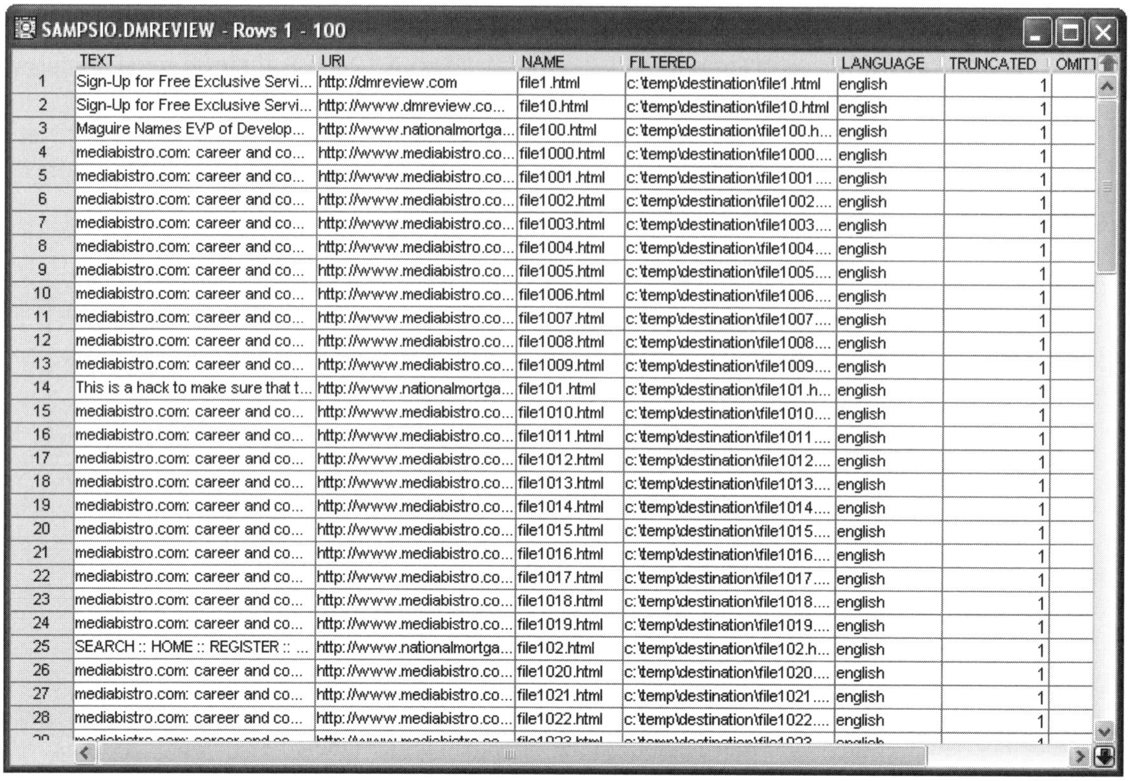

SAMPSIO.DMREVIEW - Rows 1 - 100

	TEXT	URI	NAME	FILTERED	LANGUAGE	TRUNCATED	OMITT
1	Sign-Up for Free Exclusive Servi...	http://dmreview.com	file1.html	c:\temp\destination\file1.html	english	1	
2	Sign-Up for Free Exclusive Servi...	http://www.dmreview.co...	file10.html	c:\temp\destination\file10.html	english	1	
3	Maguire Names EVP of Develop...	http://www.nationalmortga...	file100.html	c:\temp\destination\file100.h...	english	1	
4	mediabistro.com: career and co...	http://www.mediabistro.co...	file1000.html	c:\temp\destination\file1000....	english	1	
5	mediabistro.com: career and co...	http://www.mediabistro.co...	file1001.html	c:\temp\destination\file1001....	english	1	
6	mediabistro.com: career and co...	http://www.mediabistro.co...	file1002.html	c:\temp\destination\file1002....	english	1	
7	mediabistro.com: career and co...	http://www.mediabistro.co...	file1003.html	c:\temp\destination\file1003....	english	1	
8	mediabistro.com: career and co...	http://www.mediabistro.co...	file1004.html	c:\temp\destination\file1004....	english	1	
9	mediabistro.com: career and co...	http://www.mediabistro.co...	file1005.html	c:\temp\destination\file1005....	english	1	
10	mediabistro.com: career and co...	http://www.mediabistro.co...	file1006.html	c:\temp\destination\file1006....	english	1	
11	mediabistro.com: career and co...	http://www.mediabistro.co...	file1007.html	c:\temp\destination\file1007....	english	1	
12	mediabistro.com: career and co...	http://www.mediabistro.co...	file1008.html	c:\temp\destination\file1008....	english	1	
13	mediabistro.com: career and co...	http://www.mediabistro.co...	file1009.html	c:\temp\destination\file1009....	english	1	
14	This is a hack to make sure that t...	http://www.nationalmortga...	file101.html	c:\temp\destination\file101.h...	english	1	
15	mediabistro.com: career and co...	http://www.mediabistro.co...	file1010.html	c:\temp\destination\file1010....	english	1	
16	mediabistro.com: career and co...	http://www.mediabistro.co...	file1011.html	c:\temp\destination\file1011....	english	1	
17	mediabistro.com: career and co...	http://www.mediabistro.co...	file1012.html	c:\temp\destination\file1012....	english	1	
18	mediabistro.com: career and co...	http://www.mediabistro.co...	file1013.html	c:\temp\destination\file1013....	english	1	
19	mediabistro.com: career and co...	http://www.mediabistro.co...	file1014.html	c:\temp\destination\file1014....	english	1	
20	mediabistro.com: career and co...	http://www.mediabistro.co...	file1015.html	c:\temp\destination\file1015....	english	1	
21	mediabistro.com: career and co...	http://www.mediabistro.co...	file1016.html	c:\temp\destination\file1016....	english	1	
22	mediabistro.com: career and co...	http://www.mediabistro.co...	file1017.html	c:\temp\destination\file1017....	english	1	
23	mediabistro.com: career and co...	http://www.mediabistro.co...	file1018.html	c:\temp\destination\file1018....	english	1	
24	mediabistro.com: career and co...	http://www.mediabistro.co...	file1019.html	c:\temp\destination\file1019....	english	1	
25	SEARCH :: HOME :: REGISTER :: ...	http://www.nationalmortga...	file102.html	c:\temp\destination\file102.h...	english	1	
26	mediabistro.com: career and co...	http://www.mediabistro.co...	file1020.html	c:\temp\destination\file1020....	english	1	
27	mediabistro.com: career and co...	http://www.mediabistro.co...	file1021.html	c:\temp\destination\file1021....	english	1	
28	mediabistro.com: career and co...	http://www.mediabistro.co...	file1022.html	c:\temp\destination\file1022....	english	1	

Step 3: Now, in the Data Sources folder, add the DMREVIEW data source. Be sure that when you add this data set, the roles of each variable look like the ones shown in Figure 12.9.

Figure 12.9 DMREVIEW Data Set Variable Roles

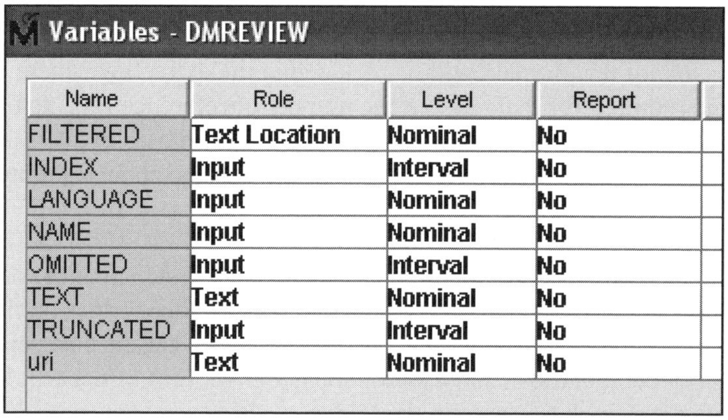

Variables - DMREVIEW

Name	Role	Level	Report
FILTERED	Text Location	Nominal	No
INDEX	Input	Interval	No
LANGUAGE	Input	Nominal	No
NAME	Input	Nominal	No
OMITTED	Input	Interval	No
TEXT	Text	Nominal	No
TRUNCATED	Input	Interval	No
uri	Text	Nominal	No

Drag the DMREVIEW data source onto your process flow and connect a Text Mining node to it. At this point, you should consider what you want in the text mining session. If you desired to mine the Web site because you wanted to find more about the kinds of topics and articles that DMReview writes for their readers, then perhaps document clustering might be beneficial in determining the topics covered in the three layers of Web pages that were extracted. If you mined the Web site because you desired to understand your competition (let's assume you were another magazine and you desired to know your competition's article depth and breadth and topics), then clustering and cluster profiles along with the concept link graphs might be useful to help uncover information about the topics.

Step 4: Let's attempt a first pass at clustering these documents and see what turns out. In the Text Mining node property sheet, set the Parsing Properties to the ones shown in Figure 12.10.

Figure 12.10 Parsing Properties

Parse	
Parse Variable	uri
Language	ENGLISH
Stop List	SASHELP.STO
Start List	
Stem Terms	Yes
Terms in Single Document	No
Punctuation	Yes
Numbers	No
Different Parts of Speech	Yes
Ignore Parts of Speech	
Noun Groups	Yes
Synonyms	SASHELP.ENO
Find Entities	No
Types of Entities	

Figure 12.11 Transform and Clustering Properties

Property	Value
Transform	
Compute SVD	Yes
SVD Resolution	Low
Max SVD Dimensions	100
Scale SVD Dimensions	No
Frequency weighting	Log
Term Weight	Entropy
Roll up Terms	No
No. of Rolled-up Terms	100
Drop Other Terms	No
Cluster	
Automatically Cluster	Yes
Exact or Maximum Number	Maximum
Number of Clusters	25
Cluster Algorithm	EXPECTATION-M
Ignore Outliers	No
Hierarchy Levels	.
Descriptive Terms	20
What to Cluster	SVD Dimensi...

We want to set the Stem Terms to Yes because term stemming is needed for any concept links to take place. It also allows SAS Text Miner to understand plural parts of speech from the root word. If you think that there are a lot of numeric representations in the documents, like financial figures and the like, then you might want to consider setting the Numbers property to Yes. In this example, we are not considering a synonym list. I will discuss more on this feature later in this chapter. Figure 12.11 shows the settings for term-document transforming and clustering. I modified the default settings to change the number of clusters to 25 and the number of descriptive terms in each cluster to 20. Run your Text Mining node. This should take about 15 minutes or so to run, depending on the machine you are using and if you are on a Windows server or a UNIX server. In my case, I was doing this on a laptop and it took about 20 minutes, give or take a few minutes. Figure 12.12 shows the Interactive results window of clusters once the Text Mining node completes. **Step 5:** To open the Interactive results window, click the Interactive property near the top of the Text Mining node property sheet.

Figure 12.12 First Pass Clustering of the DMReview Document Set

#	Descriptive Terms	Freq	Percentage ▼	RMS Std.
7	+ option, + not, + do, media professionals, + home, + post, but, personal, + call, (, community options, ,, + home, + right t, + view, free, + account, back, inc., + policy	1160	0.358578052...	5.8960305...
3	+ write, + learn, + instructor, + panel, + page, + sell, + seminar, + director, san francisco, seattle, + editor, + class, + will, + start, + location, + writer, + work, director, + letter, + online	377	0.116537867...	0.1192561...
4	+ benefit, + plan, + seminar, + book, + editor, + listing, + pitch, + magazine, + write, + member, + join, community options, media professionals, avantguild, + register, + edit, registered trademark, + log, mediabistro.com, personal	329	0.101700154...	0.1366469...
8	+ technology, data, + paper, + solution, intelligence, + issue, + application, + service, + company, + include, + resource, sourcemedia, research, + customer, + market, + business, + vendor, + provide, + product, + sale	292	0.090262751...	0.1239624...
1	exclusive, scheduled, + seminar, + sale, + resource, portals, + product, + business, + search, + covering, + issue, + vendor, + portal, dm, datawarehouse.com, + industry, + guide, + view, web, + review	272	0.084080370...	0.0424365...
5	+ trademark, + register, + email, mediabistro.com, pricing, + home, + job, copyright, + edit, info,), :, back, + member, site, + right, + career, ©, avantguild, + employer	225	0.069551777...	0.0660925...
2	+ do, + make, + good, + give, + come, + know, + work, + see, + book, + writer, + find, + year, + listing, + write, ", more, + editor, + have, + will, + business	224	0.069242658...	0.1101589...
6	+ investment, + banker, + mortgage, + market, + card, + bank, + company, + customer, data, news, |, investcorp, + subscribe, + year, + site, + product, sourcemedia, + business, privacy, + sale	179	0.055332302...	0.1034314...
9	+ contact, |, on, strictly, + conference, + advertise, + industry, + subscriber, + business, + customer, data, + online, + contain, + agreement, + service, duplication, + sale, + subscription, -, news	177	0.054714064...	0.1056920...

Notice that nine clusters have been found and the number of documents in each cluster is around 200 to 300 in size except for the cluster 7, which contains 1160 documents. If you read the descriptive keywords for each cluster, it may not be obvious what each cluster's theme is from this summary. So, we may want to take another pass and try ignoring some low information parts of speech. Figure 12.13 shows the parts of speech we will ignore for the next Text Mining node's pass at the data. **Step 6:** Reopen the Text Miner node and click Ignore parts of speech. Select the settings as shown in Figure 12.13.

Figure 12.13 Ignoring Parts of Speech in the Second Pass of Text Mining

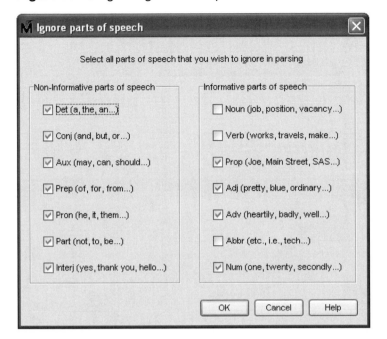

When your Text Miner node completes, open the Results window and compare the revised results in Figure 12.14 with the results in Figure 12.5. We see that Text Miner has found seven clusters, and the role by frequency plot has lessened in the number of parts of speech roles being used.
Step 7: Now, we need to profile the document clusters in order to label the themes. There is not only one correct method or only one correct set of labels to use for each cluster. I will show some examples of how to go about labeling the clusters and hope you can apply it to your specific application as well. If you open the Interactive property of the Text Mining node, the resulting clusters and key word descriptions show the seven clusters. Looking at these clusters as shown in Figure 12.15, I see that the keyword career stands out compared to the other keywords. So, moving to the Terms table, scroll down until you find the term *career*. Right-click *career* (there should be about 2200 documents indicated with this term) and select the View Concept Links option. From this concept link graph, you can see that this cluster is most likely about the careers available within this Web site, such as freelance writer, artist, etc., as you expand the sub-tree concept links. Cluster 3 appears to be mainly related to how to log on and register with Web sites to obtain e-mail newsletters and the like. Cluster 7 appears to be all about applications, systems, and solutions articles. The concept link graph for cluster 1 (research) is shown with one expanded sub-tree in Figure 12.16. I labeled the clusters in Table 12.5 from the profiling results.

Figure 12.14 Text Miner Results Window with Parts of Speech Removed

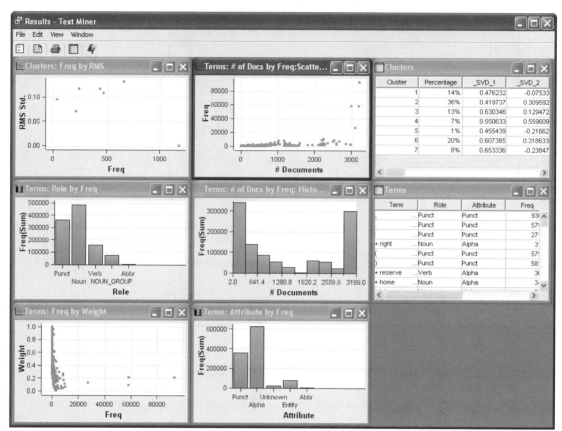

Table 12.5 Labeled Clusters from the Second Pass Text Clustering

Cluster Number	Cluster Label from Term/Concept Link Profile
2	Online community, classes, etc. Similar to Cluster 6.
6	Careers available at site, online Web courses, and registration.
1	Available research and business intelligence.
3	Site login and e-mail preferences for newsletter.
7	Applications, systems, and solutions articles.
4	Assignments and other career-related activities.
5	Online, client, and legal solutions for consumers.

Figure 12.15 Revised Clusters and Keywords with Parts of Speech Removed

#	Descriptive Terms	Freq	Percentage ▼	RMS Std.
2	+ right, + edit, + log, registered trademark, + professional, community options, + post, + click, + home, pricing, (, + call, ,, + help, media professionals, + job, + home, + employer, ©, + reserve	1174	0.36290571870...	6.9313794655213E-15
6	+ editor, + seminar, + write, + work, + benefit, + writer, + do, + book, + article, + make, + magazine, + pitch, + have, + member, + join, + career, mediabistro.com, registered trademark, wecare@mediabistro.com, + edit	657	0.20309119010...	0.13260072725557
1	+ product, + subscription, + company, + customer, + seminar, |, + contain, + business, + industry, + contact, + calendar, news, + guide, research, web, + site, + prohibit, duplication, information, + sale	467	0.14435857805...	0.10918825071146
3	login, information, + password, ., ,, + service, (,), + email, :, ©, + employer, + email, + medium, + reserve, + trademark, + right, ?, pricing, + option	436	0.13477588871...	0.11767746787548
7	+ system, + company, + vendor, + market, + investment, + application, + technology, data, + solution, management, + product, + provide, + business, + customer, + strategy, + resource, + contain, + site, + issue, + include	245	0.07573415765...	0.11706916880543
4	+ location, + editor, + month, + director, + writer, + gig, + date, + letter, + design, , + put, + sell, + write, + panel, + begin, + assignment, + class, + page, + search, + seminar	216	0.06676970633...	0.0710214625711
5	+ understand, + report, + perform, + deal, + basket, + prospect, + paper, report format, + person, + solution, + source, + consumer, + provide, + team, + paper, + help, + item, + format, + market, + vendor	40	0.01236476043...	0.0952794396256

As you can see, this process is highly interactive and requires a good deal of trial and error to find themes in clustering of textual data.

Figure 12.16 Concept Link of the Term Research in Cluster 1

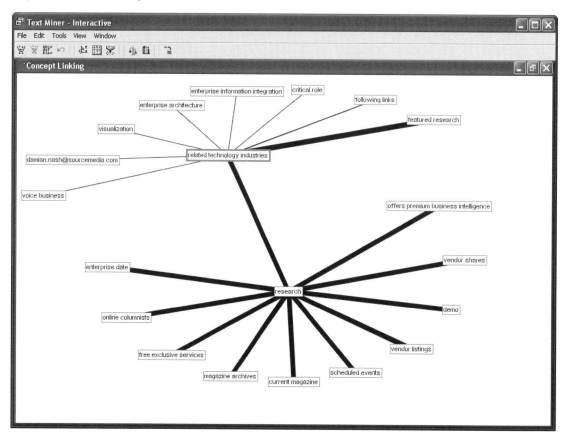

12.5 Using Text Mining in CRM Applications

It is rather difficult to demonstrate textual data in a CRM application without actually giving you data that should not be published. So, in lieu of doing this, I will describe a text application that I performed in my business and give you the thought and mining process. The business problem given to me was not really supposed to be a text mining exercise at all. I was consulting with an internal client about his sales segmentation. He had classified each of 250 or so top accounts in that industry into one of five possible groups. Four of these groups were attitudinal segments depicting the level of their technology and their feelings that this technology would help them have a competitive advantage. The fifth group was an unknown group where the marketing and sales did not know how to classify the account. When the project came to me, the marketing clients desired to know more about these segments from a product profile standpoint, which I was able to accomplish using the techniques given earlier in Chapter 10, "Product Affinity and Clustering of Product Affinities." What transpired after this profiling effort was a discussion that led to the usage of our call center database that housed unstructured notes and comments that our sales representatives had entered when communicating with customers. The desired goal in this project was to use the segmentation to create some specific campaigns with differing messaging and offers tailored to each segment group. That is when I had the idea of combining the structured and unstructured call center notes together along with the accounts that were already classified into the four segment groups. Leaving the fifth (unknown) group out of the data mining, I was able to create a predictive classification model using the unstructured notes to classify the accounts into one of the four possible attitudinal segment groups. With a classification model in hand, I could then score the fifth (unknown) accounts into one of the four segments.

Figure 12.17 Text Mining Model Account Classification Using MBR

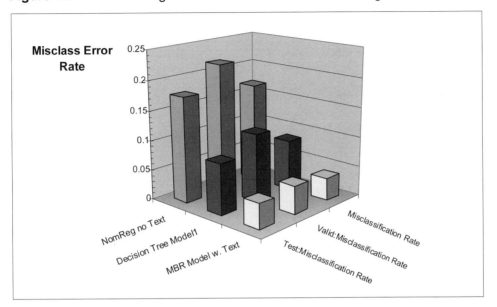

Figure 12.17 shows the classification of three different models: a regression model using only structured data, a decision tree using only structured data, and a memory-based reasoning (MBR or nearest neighbor model) of the unstructured notes data. The error rate on the MBR model using text notes was less than 5% on the training, validation, and test data sets. The MBR model was applied to the scoring of unknown accounts, and the completed scoring allowed a more complete targeted list for use in future campaigns.

12.6 References

Hearst, Marti A. 1997. "Text Data Mining." Available at
http://www.ischool.berkeley.edu/~hearst/talks/dm-talk/textfile.html.

SAS Enterprise Miner, Release 5.2. Cary, NC: SAS Institute Inc.

SAS Text Miner, Release 5.2. Cary, NC: SAS Institute Inc.

Sullivan, Dan. 2001. *Document Warehousing and Text Mining.* New York: John Wiley & Sons, Inc.

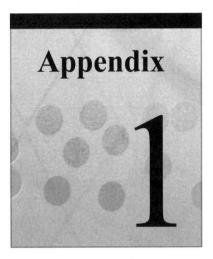

Appendix 1

Data Sets Used in Chapter Examples

Data Set BUYTEST

The variables in this data set are as follows. There are 10,000 records (rows) in this data set.

Variable Name	Type	Description
Age	Numeric	Age in years
Income	Numeric	Yearly income in thousands of dollars
Married	Numeric	(binary) 1 if married, 0 otherwise
Sex	Category	(binary) M, F
Coa6	Numeric	(binary) 1 if change of address in last 6 months, 0 otherwise
Ownhome	Numeric	(binary) 1 if own home, 0 otherwise
Loc	Category	Location of residence code: A-H
Climate	Category	Climate code for residence, 10, 20, and 30
Buy6	Numeric	Number of purchases in last 6 months
Buy12	Numeric	Number of purchases in last 12 months
Buy18	Numeric	Number of purchases in last 18 months
Value24	Numeric	Total value of purchases in past 24 months
Fico	Numeric	Credit score
Orgsrc	Category	Original customer source code: (C, D, I, O, P, R, U)
Discbuy	Numeric	(binary) 1 if a discount buyer, 0 otherwise
Return24	Numeric	(binary) 1 if product was returned in past 24 months, 0 otherwise
Respond	Numeric	(binary) 1 if responder to test mailing, 0 otherwise
Purchtot	Numeric	Test mailing purchase total
C1 – C7	Numeric	Test mailing total by product category
ID	Category	Unique ID number for each customer

Data Set CUSTOMERS

The two data sets used in the example are CUST_NEW and CUST_NEWSCORE. Both of these data sets are direct derivates of the CUSTOMERS data set with no changes in the fields.

Variable Name	Type	Description
Channel	Numeric	Channel purchase preference (0=none, 1=indirect, 2=direct, 3=both)
City	Category	City name of customer site
Corp_Rev	Numeric	Corporate revenue generated in last fiscal year incorporated
Customer	Category	(A=acquisition, R=Returning (continued purchasing), C=Churn (no purchase)
Cust_flag	Category	Empty
Cust_id	Category	Customer ID number; unique identifier for each customer
Est_spend	Numeric	Estimated Product-service spend in $ for the next full year
Loc_employee	Numeric	Number of local employees at the customer site
Prod_A	Numeric	Quantity of Product A purchased
Prod_A_Opt	Numeric	Quantity of Product A options purchased
Prod_B	Numeric	Quantity of Product B purchased
Prod_C	Numeric	Quantity of Product C purchased
Prod_D	Numeric	Quantity of Product D purchased
Prod_E	Numeric	Quantity of Product E purchased
Prod_F	Numeric	Quantity of Product F purchased
Prod_G	Numeric	Quantity of Product G purchased
Prod_H	Numeric	Quantity of Product H purchased
Prod_I	Numeric	Quantity of Product I purchased
Prod_I_Opt	Numeric	Quantity of Product I options purchased
Prod_J	Numeric	Quantity of Product J purchased
Prod_J_Opt	Numeric	Quantity of Product J options purchased
Prod_K	Numeric	Quantity of Product K purchased
Prod_L	Numeric	Quantity of Product L purchased
Prod_L_Opt	Numeric	Quantity of Product L options purchased
Prod_M	Numeric	Quantity of Product M purchased
Prod_N	Numeric	Quantity of Product N purchased
Prod_O	Numeric	Quantity of Product O purchased
Prod_O_Opt	Numeric	Quantity of Product O options purchased
Prod_P	Numeric	Quantity of Product P purchased
Prod_Q	Numeric	Quantity of Product Q purchased
Public_sector	Numeric	0 = No, 1 = Yes-public sector
Purchfst	Numeric	first year customer purchased
Purchlst	Numeric	Last year customer purchased

(continued)

Variable Name	Type	Description
Rev_class	Category	Revenue Class code; A-H (see the following explanation)
Rev_lastyr	Numeric	Last fiscal year's net operating revenues
Rev_thisyr	Numeric	This fiscal year's net operating revenues
RFM	Category	Recency, Frequency, and Monetary value cell code
Seg	Category	Industry Segment code
State	Category	State where customer site is located
Tot_Revenue	Numeric	Total net operating revenues from customer all years
US_region	Category	U.S. Region of customer location
Yrs_purchase	Numeric	Number of years customer has purchased in the past

Rev_class coding definitions

The revenue class is a coded grouping of revenue in percentile groups as follows.

A   No revenue

B   0 to 25% (bottom quartile)

C   25 to 50%

D   50 to 75%

E   75 to 90%

F   90 to 95%

G   95 to 99%

H   Top 1%

Industry Segment Code definitions

AER   Aerospace

AUT   Automotive

BKG   Banking

CHM   Chemical

CPG   Consumer Products

ELE   Electronics

FMM   Forest, Mining, and Metals

HCR   Healthcare

INS   Insurance

MED   Media and Communications

NAT   National Government

OIL   Oil and Gas

PHM   Pharmaceuticals

PSV   Professional Services

RTL   Retail/Wholesale

SLE   State/Local, and Education

TEL   Telecom

TRV   Travel

UTL   Utilities

Data Set NYTOWNS

This data set comes from Data-Miners.com (used with permission) and the column descriptions are as follows. There are 1,006 records (rows) in this data set.

NY Towns Data Column Descriptions

Variable	Type	Label
AncArab	Num	ANCESTRY (single or multiple); Total ancestries reported; Arab; Percent
AncCzech	Num	ANCESTRY (single or multiple); Total ancestries reported; Czech1; Percent
AncDanish	Num	ANCESTRY (single or multiple); Total ancestries reported; Danish; Percent
AncDutch	Num	ANCESTRY (single or multiple); Total ancestries reported; Dutch; Percent
AncEnglish	Num	ANCESTRY (single or multiple); Total ancestries reported; English; Percent
AncFrCanad	Num	ANCESTRY (single or multiple); Total ancestries reported; French Canadian1; Percent
AncFrench	Num	ANCESTRY (single or multiple); Total ancestries reported; French (except Basque)1; Percent
AncGerman	Num	ANCESTRY (single or multiple); Total ancestries reported; German; Percent
AncGreek	Num	ANCESTRY (single or multiple); Total ancestries reported; Greek; Percent
AncHungary	Num	ANCESTRY (single or multiple); Total ancestries reported; Hungarian; Percent
AncIrish	Num	ANCESTRY (single or multiple); Total ancestries reported; Irish1; Percent

(*continued*)

Variable	Type	Label
AncItalian	Num	ANCESTRY (single or multiple); Total ancestries reported; Italian; Percent
AncLithu	Num	ANCESTRY (single or multiple); Total ancestries reported; Lithuanian; Percent
AncNorweg	Num	ANCESTRY (single or multiple); Total ancestries reported; Norwegian; Percent
AncOthr	Num	ANCESTRY (single or multiple); Total ancestries reported; Other ancestries; Percent
AncPolish	Num	ANCESTRY (single or multiple); Total ancestries reported; Polish; Percent
AncPortug	Num	ANCESTRY (single or multiple); Total ancestries reported; Portuguese; Percent
AncRepP	Num	ANCESTRY (single or multiple); Total ancestries reported; Percent
AncRussian	Num	ANCESTRY (single or multiple); Total ancestries reported; Russian; Percent
AncScot	Num	ANCESTRY (single or multiple); Total ancestries reported; Scottish; Percent
AncScotIre	Num	ANCESTRY (single or multiple); Total ancestries reported; Scotch-Irish; Percent
AncSlovak	Num	ANCESTRY (single or multiple); Total ancestries reported; Slovak; Percent
AncSubSah	Num	ANCESTRY (single or multiple); Total ancestries reported; Subsaharan African; Percent
AncSwedish	Num	ANCESTRY (single or multiple); Total ancestries reported; Swedish; Percent
AncSwiss	Num	ANCESTRY (single or multiple); Total ancestries reported; Swiss; Percent
AncUS	Num	ANCESTRY (single or multiple); Total ancestries reported; United States or American; Percent
AncUkraine	Num	ANCESTRY (single or multiple); Total ancestries reported; Ukrainian; Percent
AncWIndian	Num	ANCESTRY (single or multiple); Total ancestries reported; West Indian (excluding Hispanic groups); Percent

(*continued*)

Variable	Type	Label
AncWelsh	Num	ANCESTRY (single or multiple); Total ancestries reported; Welsh; Percent
AreaLand	Num	
AreaWater	Num	
BadKitchen	Num	Occupied Housing Units; Selected characteristics; Lacking complete kitchen facilities; Percent
BadPlumbing	Num	Occupied Housing Units; Selected characteristics; Lacking complete plumbing facilities; Percent
BoatRVVan	Num	Total housing units; Units in structure; Boat, RV, van, etc.; Percent
BornAfrica	Num	Region of birth of foreign born; Total (excluding born at sea); Africa; Percent
BornAsia	Num	Region of birth of foreign born; Total (excluding born at sea); Asia; Percent
BornEurope	Num	Region of birth of foreign born; Total (excluding born at sea); Europe; Percent
BornLatAmer	Num	Region of birth of foreign born; Total (excluding born at sea); Latin America; Percent
BornNorAmer	Num	Region of birth of foreign born; Total (excluding born at sea); Northern America; Percent
BornOceania	Num	Region of birth of foreign born; Total (excluding born at sea); Oceania; Percent
BuiltAfter40	Num	Cumulative percentage of % built after X
BuiltAfter60	Num	Cumulative percentage of % built after X
BuiltAfter70	Num	Cumulative percentage of % built after X
BuiltAfter80	Num	Cumulative percentage of % built after X
BuiltAfter90	Num	Cumulative percentage of % built after X
BuiltAfter95	Num	Cumulative percentage of % built after X
BuiltAfter99	Num	Cumulative percentage of % built after X
COUNTY	Char	County
CitizenNative	Num	Nativity and place of birth; Total population; Native; Percent

(*continued*)

Variable	Type	Label
CitizenNatr	Num	Nativity and place of birth; Total population; Foreign born; Naturalized citizen; Percent
CitizenNot	Num	Nativity and place of birth; Total population; Foreign born; Not a citizen; Percent
CommuteAtHome	Num	Commuting to work; Workers 16 years and over; Worked at home; Percent
CommuteAvgTravTime	Num	Commuting to work; Workers 16 years and over; Mean travel time to work (minutes); Number
CommuteCarpool	Num	Commuting to work; Workers 16 years and over; Car, truck, or van -- carpooled; Percent
CommuteDrive	Num	Commuting to work; Workers 16 years and over; Car, truck, or van -- drove alone; Percent
CommuteOther	Num	Commuting to work; Workers 16 years and over; Other means; Percent
CommutePubTran	Num	Commuting to work; Workers 16 years and over; Public transportation (including taxicab); Percent
CommuteWalk	Num	Commuting to work; Workers 16 years and over; Walked; Percent
CostDivIncLT15	Num	Specified owner-occupied units; Selected monthly owner costs as a percentage of household income in 1999; Less than 15 percent; Percent
CostDivIncLT20	Num	Owner costs as % of household income less than 19%; Adjust for not computed
CostDivIncLT25	Num	Owner costs as % of household income less than 24%; Adjust for not computed
CostDivIncLT30	Num	Owner costs as % of household income less than 29%; Adjust for not computed
CostDivIncLT35	Num	Owner costs as % of household income less than 34% (1- % 35 or greater); Adjust for not computed
DisAdult	Num	Disability status of the civilian noninstitutionalized population; Population 21 to 64 years; With a disability; Percent
DisEmploy	Num	Disability status of the civilian noninstitutionalized population; Population 21 to 64 years; With a disability; Percent employed; Number

(*continued*)

Variable	Type	Label
DisSenior	Num	Disability status of the civilian noninstitutionalized population; Population 65 years and over; With a disability; Percent
DisYoung	Num	Disability status of the civilian noninstitutionalized population; Population 5 to 20 years; With a disability; Percent
Edu2YDeg	Num	Educational attainment; Population 25 years and over; Associate degree; Percent
Edu4YDeg	Num	Educational attainment; Population 25 years and over; Bachelors degree; Percent
EduBAplus	Num	Educational attainment; Population 25 years and over; Percent bachelors degree or higher; Number
EduColNoDeg	Num	Educational attainment; Population 25 years and over; Some college, no degree; Percent
EduGradDeg	Num	Educational attainment; Population 25 years and over; Graduate or professional degree; Percent
EduHSDip	Num	Educational attainment; Population 25 years and over; High school graduate (includes equivalency); Percent
EduHSNoDip	Num	Educational attainment; Population 25 years and over; 9th to 12th grade, no diploma; Percent
EduHSplus	Num	Educational attainment; Population 25 years and over; Percent high school graduate or higher; Number
EduLT9	Num	Educational attainment; Population 25 years and over; Less than 9th grade; Percent
FSChilPP	Num	POVERTY STATUS IN 1999 (below poverty level); Families; With related children under 18 years; With related children under 5 years; Percent below poverty level; Percent
FamIncLT100K	Num	Income in 1999; Families; Less than $100,000; Percent
FamIncLT10K	Num	Income in 1999; Families; Less than $10,000; Percent
FamIncLT150K	Num	Income in 1999; Families; Less than $150,000; Percent
FamIncLT15K	Num	Income in 1999; Families; Less than $15,000; Percent
FamIncLT200K	Num	Income in 1999; Families; Less than $200,000; Percent
FamIncLT25K	Num	Income in 1999; Families; Less than $25,000; Percent

(*continued*)

Variable	Type	Label
FamIncLT35K	Num	Income in 1999; Families; Less than $35,000; Percent
FamIncLT50K	Num	Income in 1999; Families; Less than $50,000; Percent
FamIncLT75K	Num	Income in 1999; Families; Less than $25,000; Percent
GEO_ID	Char	Geographic key from census bureau
GEO_NAME	Char	Geography
GPGuardP	Num	Grandparents as caregivers; Grandparent living in household with one or more own grandchildren under 18 years; Grandparent responsible for grandchildren; percent
HouseIncLT100K	Num	Income in 1999; Households; Less than $100,000; Percent
HouseIncLT10K	Num	Income in 1999; Households; Less than $10,000; Percent
HouseIncLT150K	Num	Income in 1999; Households; Less than $150,000; Percent
HouseIncLT15K	Num	Income in 1999; Households; Less than $15,000; Percent
HouseIncLT200K	Num	Income in 1999; Households; Less than $200,000; Percent
HouseIncLT25K	Num	Income in 1999; Households; Less than $25,000; Percent
HouseIncLT35K	Num	Income in 1999; Households; Less than $35,000; Percent
HouseIncLT50K	Num	Income in 1999; Households; Less than $50,000; Percent
HouseIncLT75K	Num	Income in 1999; Households; Less than $75,000; Percent
HouseMultiFamily	Num	% of housing units that are multifamily
IncAvgEarnings	Num	Income in 1999; Households; With earnings; Mean earnings (dollars); Number
IncAvgPubAss	Num	Income in 1999; Households; With public assistance income; Mean public assistance income (dollars); Number
IncAvgRetire	Num	Income in 1999; Households; With retirement income; Mean retirement income (dollars); Number
IncAvgSocSec	Num	Income in 1999; Households; With Social Security income; Mean Social Security income (dollars); Number
IncAvgSupSec	Num	Income in 1999; Households; With Supplemental Security Income; Mean Supplemental Security Income (dollars); Number

(continued)

Variable	Type	Label
IncEarnings	Num	Income in 1999; Households; With earnings; Percent
IncMedEarnings	Num	Income in 1999; Households; Median household income (dollars); Number
IncMedFamily	Num	Income in 1999; Families; Median family income (dollars); Number
IncMedFemales	Num	Income in 1999; Median income for Females (dollars); Number
IncMedMales	Num	Income in 1999; Median income for Males (dollars); Number
IncPerCapita	Num	Income in 1999; Families; Per capita income (dollars); Number
IncPubAss	Num	Income in 1999; Households; With public assistance income; Percent
IncRatioM2F	Num	Male to Female Median Income Ratio
IncRetirement	Num	Income in 1999; Households; With retirement income; Percent
IncSecSec	Num	Income in 1999; Households; With Social Security income; Percent
IncSupSec	Num	Income in 1999; Households; With Supplemental Security Income; Percent
IndAgric	Num	Employed civilian population 16 years and over; Industry; Agriculture, forestry, fishing and hunting, and mining; Percent
IndArtsEntertainment	Num	Employed civilian population 16 years and over; Industry; Arts, entertainment, recreation, accommodation and food services; Percent
IndConstruction	Num	Employed civilian population 16 years and over; Industry; Construction; Percent
IndEduHealthSocServ	Num	Employed civilian population 16 years and over; Industry; Educational, health and social services; Percent
IndFinanceRealEstate	Num	Employed civilian population 16 years and over; Industry; Finance, insurance, real estate, and rental and leasing; Percent
IndInformation	Num	Employed civilian population 16 years and over; Industry; Information; Percent
IndManufacturing	Num	Employed civilian population 16 years and over; Industry; Manufacturing; Percent

(continued)

Variable	Type	Label
IndOtherService	Num	Employed civilian population 16 years and over; Industry; Other services (except public administration); Percent
IndProfScienceWaste	Num	Employed civilian population 16 years and over; Industry; Professional, scientific, management, administrative, and waste management services; Percent
IndPublicAdmin	Num	Employed civilian population 16 years and over; Industry; Public administration; Percent
IndRetail	Num	Employed civilian population 16 years and over; Industry; Retail trade; Percent
IndTransport	Num	Employed civilian population 16 years and over; Industry; Transportation and warehousing, and utilities; Percent
IndWholesale	Num	Employed civilian population 16 years and over; Industry; Wholesale trade; Percent
JobAgriculture	Num	Employed civilian population 16 years and over; Occupation; Farming, fishing, and forestry occupations; Percent
JobConstruct	Num	Employed civilian population 16 years and over; Occupation; Construction, extraction, and maintenance occupations; Percent
JobManageProf	Num	Employed civilian population 16 years and over; Occupation; Management, professional, and related occupations; Percent
JobOfficeSales	Num	Employed civilian population 16 years and over; Occupation; Sales and office occupations; Percent
JobService	Num	Employed civilian population 16 years and over; Occupation; Service occupations; Percent
JobTransport	Num	Employed civilian population 16 years and over; Occupation; Production, transportation, and material moving occupations; Percent
LabCivilEmployed	Num	
LabFCivilEmployed	Num	
LaborAll	Num	Employment status; Population 16 years and over; In labor force; Percent
LaborArmedForces	Num	Employment status; Population 16 years and over; In labor force; Armed Forces; Percent
LaborCivilian	Num	Employment status; Population 16 years and over; In labor force; Civilian labor force; Percent

(*continued*)

Variable	Type	Label
LaborFemale	Num	Employment status; Females 16 years and over; In labor force; Percent
LaborUnder6ParentsWork		Num Employment status; Own children under 6 years; All parents in family in labor force; Percent
LangAsiaPoorEng	Num	Language spoken at home; Population 5 years and over; Language other than English; Asian and Pacific Island languages; Speak English less than very well; Percent
LangAsian	Num	Language spoken at home; Population 5 years and over; Language other than English; Asian and Pacific Island languages; Number
LangIEPoorEng	Num	Language spoken at home; Population 5 years and over; Language other than English; Other Indo-European languages; Speak English less than very
LangIndoEuro	Num	Language spoken at home; Population 5 years and over; Language other than English; Other Indo-European languages; Percent
LangNotEng	Num	Language spoken at home; Population 5 years and over; Language other than English; Percent
LangOnlyEng	Num	Language spoken at home; Population 5 years and over; English only; Percent
LangPoorEng	Num	Language spoken at home; Population 5 years and over; Language other than English; Speak English less than very well; Percent
LangSpan	Num	Language spoken at home; Population 5 years and over; Language other than English; Spanish; Percent
LangSpanPoorEng	Num	Language spoken at home; Population 5 years and over; Language other than English; Spanish; Speak English less than very well; Percent
Latitude	Num	Latitude of town location
LessEq1Room	Num	% with 1 room
LessEq2Rooms	Num	% with 2 or fewer rooms
LessEq3Rooms	Num	% with 3 or fewer rooms
LessEq4Rooms	Num	% with 4 or fewer rooms
LessEq5Rooms	Num	% with 5 or fewer rooms

(*continued*)

Variable	Type	Label
LessEq6Rooms	Num	% with 6 or fewer rooms
LessEq7Rooms	Num	% with 7 or fewer rooms
LessEq8Rooms	Num	% with 8 or fewer rooms (1-% with 9 or more)
Longitude	Num	Longitude of town location
MarDivorce	Num	Marital status; Population 15 years and over; Divorced; Percent
MarDivorcedFemales	Num	Marital status; Population 15 years and over; Divorced; Female; Percent
MarFemaleDivorcees	Num	% of marriage aged females who are divorced
MarFemaleWidows	Num	Percent of widows that are female
MarNever	Num	Marital status; Population 15 years and over; Never married; Percent
MarSep	Num	Marital status; Population 15 years and over; Separated; Percent
MarWidow	Num	Marital status; Population 15 years and over; Widowed; Percent
MarWidowedFemales	Num	Marital status; Population 15 years and over; Widowed; Female; Percent
Married	Num	Marital status; Population 15 years and over; Now married, except separated; Percent
MedHomeValue	Num	Specified owner-occupied units; Value; Median (dollars); Number
MedOwnerCostWMort	Num	Median Monthly ownership cost with mortgage
MedOwnerCostWOMort	Num	Median Monthly ownership cost with no mortgage
MedRoom	Num	Total housing units; Rooms; Median (rooms); Number
MedianRent	Num	Specified renter-occupied units; Gross rent; Median (dollars); Number
Mobile	Num	Total housing units; Units in structure; Mobile home; Percent
MortageLT300	Num	Percent of mortgages with payment less than $300
MortageLT500	Num	Percent of mortgages with payment less than $500
MortageLT1000	Num	Percent of mortgages with payment less than $1,000

(*continued*)

Variable	Type	Label
MortageLT1500	Num	Percent of mortgages with payment less than $1,500
MortageLT2000	Num	Percent of mortgages with payment less than $2,000
MortgageLT700	Num	Percent of mortgages with payment less than $700
NAME	Char	Area Name-Legal/Statistical Area Description (LSAD) Term-Part Indicator
NativityDifState	Num	Nativity and place of birth; Total population; Native; Born in United States; Different state; Percent
NativityForeign	Num	Nativity and place of birth; Total population; Foreign born; Percent
NativityNY	Num	Nativity and place of birth; Total population; Native; Born in United States; State of residence; Percent
NativityNewFor	Num	Nativity and place of birth; Total population; Foreign born; Entered 1990 to March 2000; Percent
NativityOutUS	Num	Nativity and place of birth; Total population; Native; Born outside United States; Percent
NativityUS	Num	Nativity and place of birth; Total population; Native; Born in United States; Percent
NoCashRent	Num	Specified renter-occupied units; Gross rent; No cash rent; Percent
NoDisEmploy	Num	Disability status of the civilian noninstitutionalized population; Population 21 to 64 years; No disability; Percent employed; Number
NoPhone	Num	Occupied Housing Units; Selected characteristics; No telephone service; Percent
OwnWMortgage	Num	Specified owner-occupied units; Mortgage status and selected monthly owner costs; With a mortgage; Percent
OwnWOMortgage	Num	Specified owner-occupied units; Mortgage status and selected monthly owner costs; Not mortgaged; Percent
Penetration	Num	Product penetration (percent of households)
PopRural	Num	Number of rural population
PopTotal	Num	Number of total population
PopUrban	Num	Number of urban population

(continued)

Variable	Type	Label
PovFamChildNoHusband	Num	POVERTY STATUS IN 1999 (below poverty level); Families with female householder, no husband present; With related children under 18 years; Number
PovFamNoHusband	Num	POVERTY STATUS IN 1999 (below poverty level); Families with female householder, no husband present; With related children under 18 years; Percent
PovFamWBaby	Num	POVERTY STATUS IN 1999 (below poverty level); Families with female householder, no husband present; Percent below poverty level; Percent
PovFamWChild	Num	POVERTY STATUS IN 1999 (below poverty level); Families; With related children under 18 years; Percent below poverty level; Percent
PovFamilies	Num	POVERTY STATUS IN 1999 (below poverty level); Families; Percent below poverty level; Percent
PovIndAdults	Num	POVERTY STATUS IN 1999 (below poverty level); Individuals; 18 years and over; Percent below poverty level; Percent
PovIndChild	Num	POVERTY STATUS IN 1999 (below poverty level); Individuals; Related children under 18 years; Percent below poverty level; Percent
PovIndSeniors	Num	POVERTY STATUS IN 1999 (below poverty level); Individuals; 18 years and over; 65 years and over; Percent below poverty level; Percent
PovIndividual	Num	POVERTY STATUS IN 1999 (below poverty level); Individuals; Percent below poverty level; Percent
RentDivIncLT15	Num	Specified renter-occupied units; Gross rent as a percentage of household income in 1999; Less than 15 percent; Percent
RentDivIncLT20	Num	Rent as % of income less than 20%; adjust for not computed
RentDivIncLT25	Num	Rent as % of income less than 25%; adjust for not computed
RentDivIncLT30	Num	Rent as % of income less than 30%; adjust for not computed
RentDivIncLT35	Num	Rent as % of income less than 35% (1- % 35 or over); adjust for not computed
RentLT200	Num	% rent less than $200
RentLT300	Num	% rent less than $300

(*continued*)

Variable	Type	Label
RentLT500	Num	% rent less than$ 500
RentLT750	Num	% rent less than $750
RentLT1000	Num	% rent less than $1,000
RentLT1500	Num	% rent less than $1,500
ResDfCnty	Num	Residence in 1995; Population 5 years and over; Different house in the U.S. in 1995; Different county; Percent
ResDfHome	Num	Residence in 1995; Population 5 years and over; Different house in the U.S. in 1995; Percent
ResDfState	Num	Residence in 1995; Population 5 years and over; Different house in the U.S. in 1995; Different county; Different state; Percent
ResElse	Num	Residence in 1995; Population 5 years and over; Elsewhere in 1995; Percent
ResSame	Num	Residence in 1995; Population 5 years and over; Same house in 1995; Percent
ResSmCnty	Num	Residence in 1995; Population 5 years and over; Different house in the U.S. in 1995; Same county; Percent
ResSmState	Num	Residence in 1995; Population 5 years and over; Different house in the U.S. in 1995; Different county; Same state; Percent
Resid10Yrs	Num	Resident for 10 or more years
Resid1Yr	Num	Resident for 1 or more years
Resid20Yrs	Num	Resident for 20 or more years
Resid30Yrs	Num	Resident for 30 or more years
Resid5Yrs	Num	Resident for 5 or more years
SchElem	Num	School enrollment; Population 3 years and over enrolled in school; Elementary school (grades 1-8); Percent
SchHS	Num	School enrollment; Population 3 years and over enrolled in school; High school (grades 9-12); Percent
SchKind	Num	School enrollment; Population 3 years and over enrolled in school; Kindergarten; Percent
SchNurs	Num	School enrollment; Population 3 years and over enrolled in school; Nursery school, preschool; Percent

(continued)

Variable	Type	Label
SchUniv	Num	School enrollment; Population 3 years and over enrolled in school; College or graduate school; Percent
ValueLT100K	Num	% value less than $100,000
ValueLT150K	Num	% value less than $150,000
ValueLT1Mil	Num	% value less than $1,000,000 (1-% over $1,000,000)
ValueLT200K	Num	% value less than $200,000
ValueLT300K	Num	% value less than $300,000
ValueLT500K	Num	% value less than $500,000
ValueLT50K	Num	% value less than $50,000
VehicGT0	Num	At least one vehicle available
VehicGT1	Num	More than one vehicle available
VehicGT2	Num	More than two vehicles available
VeteranC	Num	Veteran status; Civilian population 18 years and over; Civilian veterans; Percent
WorkClassFamily	Num	Employed civilian population 16 years and over; Class of worker; Unpaid family workers; Percent
WorkClassGovt	Num	Employed civilian population 16 years and over; Class of worker; Government workers; Percent
WorkClassSalary	Num	Employed civilian population 16 years and over; Class of worker; Private/wage & salary workers; Percent
WorkClassSelf	Num	Employed civilian population 16 years and over; Class of worker; Self-employed workers in own not incorporated business; Percent

Data Set NEWS

This data set contains 600 records (rows).

Variable Name	Type	Description
TEXT	Text	Contains the text of a sample of news stories.
Graphics	Numeric	Binary (0 and 1) – 1 means the text is about graphics; 0 means not about graphics.
Hockey	Numeric	Binary (0 and 1) – 1 means the text is about hockey; 0 means not about hockey.
Medical	Numeric	Binary (0 and 1) – 1 meaning the text is about a medical topic; 0 means not about a medical topic.

The data for the Web crawling is derived from DMReview.com Web site and the content changes over time.

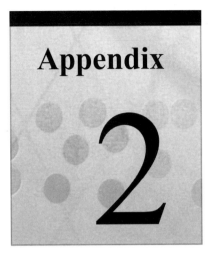

Appendix 2

Contents of the Data CD

The data CD contains folders for Chapters 2 through 12. For their respective chapters, these folders provide the data sets used in the examples, and the XML files, which are the process flow diagrams of the completed exercises. A folder is not listed for Chapter 1 because this chapter does not contain data sets or process flow diagrams.

Chapter 2:

Name	Ext	Size	Type
🔼			
buytest.sas7bdat	.sa...	1,868,800	SAS D...
Data Assay.xml	.xml	3,986	XML D...

Chapter 3:

Name	Ext	Size	Type
🔼			
distance.gif	.gif	2,579	GIF Ima...
distance_example.sas	.sas	967	SAS Pr...
more_detail_distance_ex.sas	.sas	3,281	SAS Pr...
tree.gif	.gif	3,590	GIF Ima...

Chapter 4:

Name	Ext	Size	Type
🔼			
RFM Cells.xml	.xml	15,168	XML Docu...
rfm.sas	.sas	1,269	SAS Program
rfm_format.sas	.sas	519	SAS Program

Chapter 5:

Name	Ext	Size	Type
⬆...			
SAS Code Process Flow Table2.sas	.sas	302	SAS Program
B-B Segmentation.xml	.xml	12,122	XML Document
Decision Tree Clustering.xml	.xml	22,692	XML Document
customers.sas7bdat	.sas7...	36,783,104	SAS Data Set
revised_customer.sas7bdat	.sas7...	36,783,104	SAS Data Set

Chapter 6:

Name	Ext	Size	Type
⬆...			
NY Towns Clustering.xml	.xml	21,437	XML Docu...
nytowns.sas7bdat	.sa...	2,409,472	SAS Data ...
Softmax.sas	.sas	1,667	SAS Program

Chapter 7:

Name	Ext	Size	Type
⬆...			
cust_new.sas7bdat	.sa...	10,896,384	SAS Data ...
cust_newscore.sas7bdat	.sa...	1,115,136	SAS Data ...
cust_subset.sas7bdat	.sa...	25,723,904	SAS Data ...
Distance Detection.xml	.xml	17,248	XML Docu...

Chapter 8:

Name	Ext	Size	Type
⬆...			
Buyer Segmentation.xml	.xml	10,471	XML Docu...
customer_value.sas7bdat	.sa...	3,572,736	SAS Data ...
MVC Process Flow.xml	.xml	9,794	XML Docu...

Chapter 9:

Name	Ext	Size	Type
Clustering with missing data.sas	.sas	3,368	SAS Program
County Clusters.xml	.xml	19,977	XML Docu...
Effects of Missing Data on Cl...	.xml	18,679	XML Docu...
jackboot.sas	.sas	37,318	SAS Program
Missing Data Pattern.xml	.xml	2,315	XML Docu...
missing_patterns.htm	.htm	26,514	HTML Doc...
nytowns_imputed.sas7bdat	.sa...	2,409,472	SAS Data ...
nytowns_wmissing.sas7bdat	.sa...	2,409,472	SAS Data ...
proc_mi.sas	.sas	3,334	SAS Program

Chapter 10:

Name	Ext	Size	Type
Affinity by Segment.sas	.sas	790	SAS Program
Binary Product Affinity.xml	.xml	48,056	XML Document
Compute Binary Product Affinities.sas	.sas	1,942	SAS Program
customer_score.sas7bdat	.sas7...	45,483,008	SAS Data Set
Exercise 1 answer.sas	.sas	1,298	SAS Program
Graph Theory Approach.xml	.xml	10,521	XML Document
soft_score.sas7bdat	.sas7...	64,013,312	SAS Data Set
Softmax_transform.sas	.sas	2,129	SAS Program

Chapter 11:

Name	Ext	Size	Type
Customer Distinction.xml	.xml	13,753	XML Document
rfm_score_test.sas7bdat	.sas7...	377,856	SAS Data Set
softmax scoring.sas	.sas	3,151	SAS Program
SOM Fuzzy Segments.xml	.xml	15,755	XML Document
som_data.sas7bdat	.sas7...	50,840,576	SAS Data Set
survey.sas	.sas	3,284	SAS Program

Chapter 12:

Name	Ext	Size	Type
News Stories.xml	.xml	8,858	XML Docu...
news.sas7bdat	.sa...	20,002,304	SAS Data ...
Text Clustering.xml	.xml	2,192	XML Docu...

270

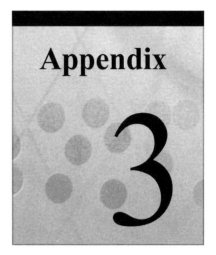

Appendix

3

SAS Programs and Macros on the Data CD

Chapter 3 Programs

The two following SAS programs compute distance metrics:

Distance_example.sas

```
data work.age_value;
   input cust_id $ age value;
   cards;
Cust_A  8 18.50
Cust_B  3  3.30
Cust_C  1  9.75
;
run;

title 'Distances of Just the AGE Variable';
proc distance data=age_value out=age_dist method=Euclid;
   var interval(age);
id cust_id;
run;
title;

title 'Compute the Distances of AGE & VALUE Combined';
proc distance data=age_value out=Dist method=Euclid;
     var interval(age value);
     id cust_id;
   run;
title;

proc mds data=dist level=absolute out=mds_out;
id cust_id;
run;
%let plotitop =  gopts   = gsfmode = replace
                    device = gif
                    gaccess = gsasfile
                    hsize  = 5.63    vsize  = 3.5
                    cback  = white,
                    color  = black,
```

```
                      colors  = black,
                      options = noclip border expand,
post=c:\temp\distance.gif;

%plotit(data=mds_out,datatype=mds, labelvar=cust_id, vtoh=1.75);
```

more_detail_distance_ex.sas

```
data work.simple_cust;
   input cust_id $ age revenue state $;

   cards;
372185321         28        155      CA
075189457         55        68       WA
538590043         32        164      OH
112785896         40        596      PA
678408574         26        48       ME
009873687         45        320      KS
138569322         37        190      FL
;
run;

data work.simple_customer;
   set work.simple_cust;
     if state='CA' then state_ca=1; else state_ca=0;
    if state='OH' then state_oh=1; else state_oh=0;
    if state='PA' then state_pa=1; else state_pa=0;
    if state='ME' then state_me=1; else state_me=0;
    if state='KS' then state_ks=1; else state_ks=0;
    if state='FL' then state_fl=1; else state_fl=0;
    if state='WA' then state_wa=1; else state_wa=0;
run;

title 'Distances of AGE Variable';
proc distance data=simple_customer out=age_dist method=Euclid;
  var interval(age revenue /std=std) interval (state_ca state_oh
state_pa state_me
    state_ks state_fl state_wa /std=std);
id cust_id;
run;
title;


proc mds data=age_dist level=absolute out=mds_out;
id cust_id;
run;
%let plotitop =  gopts   = gsfmode = replace
                    device = win
                    cback   = white,
                    color   = black,
                    colors  = black,
                    options = noclip border expand;

%plotit(data=mds_out,datatype=mds, labelvar=cust_id, vtoh=1.75);

proc cluster data=simple_customer outtree=Tree method=ward
                ccc pseudo print=15;
     var age revenue state_ca state_oh state_pa state_me state_ks
    state_fl state_wa;
      id Cust_id;
   run;
```

```
goptions reset=all;
goptions axis1 order=(0 to 1 by 0.2);
   proc tree data=Tree out=New nclusters=6
             graphics haxis=axis1 horizontal;
      height _rsq_    ;
   copy age revenue state_ca state_oh state_pa state_me state_ks
      state_fl state_wa;
        id cust_id;
   run;
```

Chapter 4 Programs

This code is used to generate RFM cells in Figure 4.5:

```
/* RFM Cell code development - Chapt. 4 */
data  &EM_EXPORT_TRAIN;
length rfm $1;
   set &EM_IMPORT_DATA;
     if (PCTL_VALUE24)='01:low -149' then do;
           if buy18=0 and buy12=0 and buy6=0 then RFM='A';
           if buy18 ge 1 or buy12 ge 1 or buy6 ge 1 then RFM='B';
            if buy6=1 and buy12=1 and buy18=1 then RFM='C';
        end;
       if (PCTL_VALUE24)='02:149-214' then do;
           if buy18=0 and buy12=0 and buy6=0 then RFM='D';
           if buy18 ge 1 or buy12 ge 1 or buy6 ge 1 then RFM='E';
            if buy6=1 and buy12=1 and buy18=1 then RFM='F';
        end;
       if (PCTL_VALUE24)='03:214-312' then do;
           if buy18=0 and buy12=0 and buy6=0 then RFM='G';
           if buy18 ge 1 or buy12 ge 1 or buy6 ge 1 then RFM='H';
            if buy6=1 and buy12=1 and buy18=1 then RFM='I';
        end;
       if (PCTl_VALUE24)='04:312-high' then do;
               if buy18=0 and buy12=0 and buy6=0 then RFM='J';
           if buy18 ge 1 or buy12 ge 1 or buy6 ge 1 then RFM='K';
            if buy6=1 and buy12=1 and buy18=1 then RFM='L';
        end;
run;

options ls=80 ps=50 nodate nonumber;
title 'Distribution of RFM Cells on BUYTEST data';
proc freq data=&EM_EXPORT_TRAIN;
   table rfm;
   run;
title;
```

RFM_format.sas

```
proc format;
  value $rfm
   A = 'A: Bottom 25%, No Purch 18mo'
   B = 'B: Bottom 25%, Purch within 18mo'
   C = 'C: Bottom 25%, Purch within 6-12mo'
   D = 'D: Middle 50%, No Purch 18mo'
   E = 'E: Middle 50%, Purch within 18mo'
   F = 'F: Middle 50%, Purch within 6-12mo'
   G = 'G: Upper 25%, No Purch 18mo'
   H = 'H: Upper 25%, Purch within 18mo'
```

```
      I = 'I: Upper 25%, Purch within 6-12mo'
      J = 'J: Top 25%, No Purch 18mo'
      K = 'K: Top 25%, Purch within 18mo'
      L = 'L: Top 25%, Purch within 6-12mo';
   run;
```

Chapter 6 Programs

This macro computes the softmax function:

Softmax.sas

```
/* Macro Softmax Transform                                    */
/* This macro computes one of three scaling transforms, linear, */
/* unscaled softmax, and scaled or squashed softmax.           */
/* The confidence level used is 90% which is a normal z score of */
/* about 1.283 which is fixed in this application.             */
/* log=L for linear scale, log=S for unscaled softmax, log=SS for */
/* squashed-scaled softmax.                                     */
%macro softmax(dsin=,var=,dsout=,log=L);

 proc sql;
   create table work.stats as
  select min(&var) as minv,
         max(&var) as maxv,
         mean(&var) as meanv,
         std(&var) as stdev
    from &dsin
 ;
quit;
   data _null_ ;
     set work.stats;
       call symput('minv',minv);
       call symput('maxv',maxv);
       call symput('meanv',meanv);
       call symput('stdev',stdev);
   run;
%if %upcase(&log)=L %then
    %do;
        data &dsout;
        set &dsin;
         sm_&var = (&var - &minv)/(&maxv - &minv);
        run;
    %end;

  %else
    %if %upcase(&log)=S %then
       %do;
           data &dsout;
              set &dsin;
              %let var1 = (&var - &meanv)/(1.283 *
(&stdev/6.2831853));
              sm_&var = 1/(1 + exp(- &var1));
          run;
       %end;
  %else
```

```
    %if %upcase(&log)=SS %then
        %do;
            data &dsout;
                set &dsin;
                  %let var1 = (&var - &meanv)/(1.283 *
(&stdev/6.2831853));
                  %let var2 = 1/(1 + exp(- &var1));
                    sm_&var = &meanv +
                    ((1.283 * &stdev*log(sqrt(-1+(1/(1-&var2))))) )/
3.14159265 );
              run;
        %end;
%mend softmax;
```

Chapter 9 Programs

The following program imputes missing entries along with macro statements for SAS Enterprise
Miner reporting macro calls:

Proc MI.sas

```
options ls=132 ps=50 nodate nonumber;
/* Take scored cluster results from County Clusters diagram and sort
by cluster*/
/* segment */
proc sort data=sampsio.nytowns_county;
 by _segment_ ;
 run;
ods html
body='c:\temp\proc_mi.htm' style=BarrettsBlue;
title 'Missing Imputation Analysis';
 proc mi data=sampsio.nytowns_county
    seed=1234567 nimpute=5 minmaxiter=200 min=. . . . 0    6    0
                                          max=. . . . 48.5 59.5 44.8
    out=work.total_impute ;
mcmc chain=single displayinit initial=em(itprint) impute=full
outest=work.mcmc_est ;
  by _segment_ ;
var mortgagelt700 valuelt50k indagric marfemaledivorcees
    penetration eduhsdip ancitalian;
run;
ods html close;
proc summary data=work.total_impute mean nway ;
 class geo_id;
var penetration eduhsdip ancitalian ;
output out=work.impute_means mean= ;
  where missing_flag=1;
run;
proc sort data=work.impute_means ;
  by geo_id; run;
proc sort data=sampsio.nytowns_county out=work.orig_ds;
by geo_id;
run;
data sampsio.nytowns_imputed;
  merge work.orig_ds(in=a)
  work.impute_means(in=b drop=_freq_);
by geo_id;
  if a;
run;
proc means data=sampsio.nytowns mean stderr noprint;
  var penetration eduhsdip ancitalian ;
```

```
output out=work.orig_mean mean= stderr=stderr(penetration)=stde_penet
        stderr=stderr(eduhsdip)=stde_eduh
stderr=stderr(ancitalian)=stde_anci ;
run;
proc means data=sampsio.nytowns_imputed mean stderr noprint;
  var penetration eduhsdip ancitalian ;
output out=work.impute_mean mean=
stderr=stderr(penetration)=stde_penet
        stderr=stderr(eduhsdip)=stde_eduh
stderr=stderr(ancitalian)=stde_anci ;
run;

/* Now format the output of the Mean-StdErr data set */
ods html
body='c:\temp\mi_compare.htm' style=BarrettsBlue;
title 'Imputed Means & Std.Errors on Imputed Data set';
proc sql;
  select _freq_ label='# of Obs',
      penetration label='Mean Penetration',
    eduhsdip label='Mean Edu Attainment',
    ancitalian label='Mean Italian Ancestry',
    stde_penet label='SE Penetration',
    stde_eduh label='SE Edu Attainment',
    stde_anci label='SE Italian Ancestry'
    from work.impute_mean
;
quit;
title 'NYTowns Means & Std.Errors on Original Data Set';
proc sql;
  select _freq_ label='# of Obs',
      penetration label='Mean Penetration',
    eduhsdip label='Mean Edu Attainment',
    ancitalian label='Mean Italian Ancestry',
    stde_penet label='SE Penetration',
    stde_eduh label='SE Edu Attainment',
    stde_anci label='SE Italian Ancestry'
    from work.orig_mean
;
quit;
ods html close;
%em_register(key=first,type=data);
%em_register(key=second,type=data);
%em_register(key=third,type=data);
data &em_user_first;
  set emws.clus_train;
run;
data &em_user_second;
 set emws.clus4_train;
run;
data &em_user_third;
  set emws.clus5_train; run;
proc sort data=first; by _segment_ ;
proc sort data=second; by _segment_;
proc sort data=third; by _segment_;
run;
%em_report(key=first,viewtype=histogram,x=distance,block=Plots,
by=_segment_,description=Orig Clustering Distances);
%em_report(key=second,viewtype=histogram,x=distance,block=Plots,
by=_segment_,description=Mean Imputation Cluster Distances);
%em_report(key=third,viewtype=histogram,x=distance,block=Plots,
by=_segment_,description=MI Imputed Cluster Distances);
```

Chapter 10 Programs

The following program computes the product affinity scores for each cluster segment:

Affinity by segment.sas

```
/* Calculating overall and cluster means    */
ods html style=barrettsblue body='c:\temp\bin_qty_means.htm';
title 'Product Quantity Affinity by Segment Means';
proc means data=EMWS.META_TRAIN mean ;
  class _segment_ ;
 var prod_a prod_b prod_c prod_d prod_e prod_f prod_g
     prod_h prod_i prod_j prod_k prod_l prod_m prod_n prod_o
     prod_p prod_q prod_a_opt prod_b_opt prod_e_opt prod_i_opt
     prod_j_opt prod_o_opt prod_l_opt ;
run;

title 'Product Binary Affinity by Segment Means';
proc means data=EMWS.META_TRAIN mean ;
  class _segment_ ;
 var bin_a bin_b bin_c bin_d bin_e bin_f bin_g
     bin_h bin_i bin_j bin_k bin_l bin_m bin_n bin_o
     bin_p bin_q bin_a_opt bin_b_opt bin_e_opt bin_i_opt
     bin_j_opt bin_o_opt bin_l_opt ;
run;
ods html close;
```

The binary product affinity is given in:

```
Compute Binary Product Affinity.sas

data &EM_EXPORT_TRAIN;
 set &EM_IMPORT_DATA;
 if prod_a=. then prod_a=0;  if prod_a_opt=. then prod_a_opt=0;
 if prod_b=. then prod_b=0;  if prod_b_opt=. then prod_b_opt=0;
 if prod_c=. then prod_c=0;
 if prod_d=. then prod_d=0;
 if prod_e=. then prod_e=0; if prod_e_opt=. then prod_e_opt=0;
 if prod_f=. then prod_f=0;
 if prod_g=. then prod_g=0;
 if prod_h=. then prod_h=0;
 if prod_i=. then prod_i=0; if prod_i_opt=. then prod_i_opt=0;
 if prod_j=. then prod_j=0; if prod_j_opt=. then prod_j_opt=0;
 if prod_k=. then prod_k=0;
 if prod_l=. then prod_l=0; if prod_l_opt=. then prod_l_opt=0;
 if prod_m=. then prod_m=0;
 if prod_n=. then prod_n=0;
 if prod_o=. then prod_o=0; if prod_o_opt=. then prod_o_opt=0;
 if prod_p=. then prod_p=0;
 if prod_q=. then prod_q=0;
 /* Compute binary product scores from quantities */
   if prod_a=0 then bin_a=0; else bin_a=1;
   if prod_a_opt=0 then bin_a_opt=0; else bin_a_opt=1;
   if prod_b=0 then bin_b=0; else bin_b=1;
   if prod_b_opt=0 then bin_b_opt=0; else bin_b_opt=1;
   if prod_c=0 then bin_c=0; else bin_c=1;
   if prod_d=0 then bin_d=0; else bin_d=1;
   if prod_e=0 then bin_e=0; else bin_e=1;
   if prod_e_opt=0 then bin_e_opt=0; else bin_e_opt=1;
   if prod_f=0 then bin_f=0; else bin_f=1;
   if prod_g=0 then bin_g=0; else bin_g=1;
   if prod_h=0 then bin_h=0; else bin_h=1;
```

```
      if prod_i=0 then bin_i=0; else bin_i=1;
      if prod_i_opt=0 then bin_i_opt=0; else bin_i_opt=1;
      if prod_j=0 then bin_j=0; else bin_j=1;
      if prod_j_opt=0 then bin_j_opt=0; else bin_j_opt=1;
      if prod_k=0 then bin_k=0; else bin_k=1;
      if prod_l=0 then bin_l=0; else bin_l=1;
      if prod_l_opt=0 then bin_l_opt=0; else bin_l_opt=1;
      if prod_m=0 then bin_m=0; else bin_m=1;
      if prod_n=0 then bin_n=0; esle bin_n=1;
      if prod_o=0 then bin_o=0; else bin_o=1;
      if prod_o_opt=0 then bin_o_opt=0; else bin_o_opt=1;
      if prod_p=0 then bin_p=0; else bin_p=1;
      if prod_q=0 then bin_q=0; else bin_q=1;
 run;

 data &EM_EXPORT_TRAIN;
  set &EM_IMPORT_DATA;
  if prod_a=. then prod_a=0;   if prod_a_opt=. then prod_a_opt=0;
  if prod_b=. then prod_b=0;   if prod_b_opt=. then prod_b_opt=0;
  if prod_c=. then prod_c=0;
  if prod_d=. then prod_d=0;
  if prod_e=. then prod_e=0;  if prod_e_opt=. then prod_e_opt=0;
  if prod_f=. then prod_f=0;
  if prod_g=. then prod_g=0;
  if prod_h=. then prod_h=0;
  if prod_i=. then prod_i=0;  if prod_i_opt=. then prod_i_opt=0;
  if prod_j=. then prod_j=0;  if prod_j_opt=. then prod_j_opt=0;
  if prod_k=. then prod_k=0;
  if prod_l=. then prod_l=0;  if prod_l_opt=. then prod_l_opt=0;
  if prod_m=. then prod_m=0;
  if prod_n=. then prod_n=0;
  if prod_o=. then prod_o=0;  if prod_o_opt=. then prod_o_opt=0;
  if prod_p=. then prod_p=0;
  if prod_q=. then prod_q=0;
 /* Compute binary product scores from quantities */
      if prod_a=0 then bin_a=0; else bin_a=1;
      if prod_a_opt=0 then bin_a_opt=0; else bin_a_opt=1;
      if prod_b=0 then bin_b=0; else bin_b=1;
      if prod_b_opt=0 then bin_b_opt=0; else bin_b_opt=1;
      if prod_c=0 then bin_c=0; else bin_c=1;
      if prod_d=0 then bin_d=0; else bin_d=1;
      if prod_e=0 then bin_e=0; else bin_e=1;
      if prod_e_opt=0 then bin_e_opt=0; else bin_e_opt=1;
      if prod_f=0 then bin_f=0; else bin_f=1;
      if prod_g=0 then bin_g=0; else bin_g=1;
      if prod_h=0 then bin_h=0; else bin_h=1;
      if prod_i=0 then bin_i=0; else bin_i=1;
      if prod_i_opt=0 then bin_i_opt=0; else bin_i_opt=1;
      if prod_j=0 then bin_j=0; else bin_j=1;
      if prod_j_opt=0 then bin_j_opt=0; else bin_j_opt=1;
      if prod_k=0 then bin_k=0; else bin_k=1;
      if prod_l=0 then bin_l=0; else bin_l=1;
      if prod_l_opt=0 then bin_l_opt=0; else bin_l_opt=1;
      if prod_m=0 then bin_m=0; else bin_m=1;
      if prod_n=0 then bin_n=0; else bin_n=1;
      if prod_o=0 then bin_o=0; else bin_o=1;
      if prod_o_opt=0 then bin_o_opt=0; else bin_o_opt=1;
      if prod_p=0 then bin_p=0; else bin_p=1;
      if prod_q=0 then bin_q=0; else bin_q=1;
 run;
```

Chapter 11 Programs

The following program computes the softmax function for scoring:

Softmax scoring.sas

```
/* Macro Softmax Transform                                    */
/* This macro computes one of three scaling transforms, linear, */
/* unscaled softmax, and scaled or squashed softmax.            */
/* The confidence level used is 90% which is a normal z score of */
/* about 1.283 which is fixed in this application.              */
/* log=L for linear scale, log=S for unscaled softmax, log=SS for */
/* squashed-scaled softmax.                                    */
%macro softmax(dsin=,var=,dsout=,log=L);

 proc sql;
   create table work.stats as
  select min(&var) as minv,
         max(&var) as maxv,
         mean(&var) as meanv,
         std(&var) as stdev
    from &dsin
 ;
quit;
    data _null_ ;
      set work.stats;
        call symput('minv',minv);
        call symput('maxv',maxv);
        call symput('meanv',meanv);
        call symput('stdev',stdev);
    run;
%if %upcase(&log)=L %then
     %do;
         data &dsout;
         set &dsin;
          sm_&var = (&var - &minv)/(&maxv - &minv);
         run;
      %end;

  %else
    %if %upcase(&log)=S %then
        %do;
            data &dsout;
                set &dsin;
                %let var1 = (&var - &meanv)/(1.283 *
(&stdev/6.2831853));
                sm_&var = 1/(1 + exp(- &var1));
            run;
        %end;
  %else
    %if %upcase(&log)=SS %then
        %do;
            data &dsout;
                set &dsin;
                 %let var1 = (&var - &meanv)/(1.283 *
(&stdev/6.2831853));
                %let var2 = 1/(1 + exp(- &var1));
                    sm_&var = &meanv +
                    ((1.283 * &stdev*log(sqrt(-1+(1/(1-&var2))))) )/
3.14159265 );
            run;
        %end;
%mend softmax;
```

```
data work.test; set &em_import_data;
  if prod_a=. then prod_a=0;
  if prod_b=. then prod_b=0;
  if prod_c=. then prod_c=0;
  if prod_d=. then prod_d=0;
  if prod_e=. then prod_e=0;
  if prod_f=. then prod_f=0;
  if prod_g=. then prod_g=0;
  if prod_h=. then prod_h=0;
  if prod_i=. then prod_i=0;
  if prod_j=. then prod_j=0;
  if prod_k=. then prod_k=0;
  if prod_l=. then prod_l=0;
  if prod_m=. then prod_m=0;
  if prod_n=. then prod_n=0;
  if prod_o=. then prod_o=0;
  if prod_p=. then prod_p=0;
  if prod_q=. then prod_q=0;
run;

%softmax(dsin=work.test,dsout=work.temp,var=prod_a,log=S);
%softmax(dsin=work.temp,dsout=work.temp,var=prod_b,log=S);
%softmax(dsin=work.temp,dsout=work.temp,var=prod_c,log=S);
%softmax(dsin=work.temp,dsout=work.temp,var=prod_d,log=S);
%softmax(dsin=work.temp,dsout=work.temp,var=prod_e,log=S);
%softmax(dsin=work.temp,dsout=work.temp,var=prod_f,log=S);
%softmax(dsin=work.temp,dsout=work.temp,var=prod_g,log=S);
%softmax(dsin=work.temp,dsout=work.temp,var=prod_h,log=S);
%softmax(dsin=work.temp,dsout=work.temp,var=prod_i,log=S);
%softmax(dsin=work.temp,dsout=work.temp,var=prod_j,log=S);
%softmax(dsin=work.temp,dsout=work.temp,var=prod_k,log=S);
%softmax(dsin=work.temp,dsout=work.temp,var=prod_l,log=S);
%softmax(dsin=work.temp,dsout=work.temp,var=prod_m,log=S);
%softmax(dsin=work.temp,dsout=work.temp,var=prod_n,log=S);
%softmax(dsin=work.temp,dsout=work.temp,var=prod_o,log=S);
%softmax(dsin=work.temp,dsout=work.temp,var=prod_p,log=S);
%softmax(dsin=work.temp,dsout=sampsio.som_data,var=prod_q,log=S);
```

The following program is used to merge in survey data with the sample of customer data for further analysis.

Survey.sas

```
%macro survey;
/* Question formats:  */
/* Q1: Do you purchase typically from paper Catalogs? */
/* Q2: Do you like receiving offers based on your category selection? */
/* Q3: Do you purchase from a catalog more than once per year? */
/* Q4: Do you also purchase from on-line web sites? */
/* Q5: Did you fill out a profile that was in our last catalog? */

data &em_export_train;
  set &em_import_data;
 if respond=1 then do;
        q1 = 1;
        q2 = 1;
        q3 = 1;
        q4 = 0;
        q5 = 1;
    end;
```

```
  if discbuy=1 then do;
     q1 = 1;
        q2 = 1;
        q3 = 0;
        q4 = 1;
        q5 = 0;
 end;
 if married=1 and fico < 700 then do;
     q1 = 1;
        q2 = 0;
        q3 = 0;
        q4 = 1;
        q5 = 1;
 end;
    if q1=. or q2=. or q3=. or q4=. or q5=. then
 do;
      q1=0;
      q2=0;
      q3=0;
      q4=0;
      q5=0;
 end;
end;

survey_score = sum(q1,q2,q3,q4,q5);
run;
quit;
%mend survey;

%survey;
```

Index

Books Available from SAS Press

Advanced Log-Linear Models Using SAS®
by **Daniel Zelterman**

Analysis of Clinical Trials Using SAS®: A Practical Guide
by **Alex Dmitrienko, Geert Molenberghs, Walter Offen,** *and*
Christy Chuang-Stein

Annotate: Simply the Basics
by **Art Carpenter**

*Applied Multivariate Statistics with SAS® Software,
Second Edition*
by **Ravindra Khattree**
and **Dayanand N. Naik**

*Applied Statistics and the SAS® Programming Language,
Fifth Edition*
by **Ronald P. Cody**
and **Jeffrey K. Smith**

An Array of Challenges — Test Your SAS® Skills
by **Robert Virgile**

*Building Web Applications with SAS/IntrNet®: A Guide to the
Application Dispatcher*
by **Don Henderson**

*Carpenter's Complete Guide to the SAS® Macro Language,
Second Edition*
by **Art Carpenter**

Carpenter's Complete Guide to the SAS® REPORT Procedure
by **Art Carpenter**

The Cartoon Guide to Statistics
by **Larry Gonick**
and **Woollcott Smith**

*Categorical Data Analysis Using the SAS® System,
Second Edition*
by **Maura E. Stokes, Charles S. Davis,**
and **Gary G. Koch**

Cody's Data Cleaning Techniques Using SAS® Software
by **Ron Cody**

*Common Statistical Methods for Clinical Research with
SAS® Examples, Second Edition*
by **Glenn A. Walker**

The Complete Guide to SAS® Indexes
by **Michael A. Raithel**

*Data Management and Reporting Made Easy with
SAS® Learning Edition 2.0*
by **Sunil K. Gupta**

Data Preparation for Analytics Using SAS®
by **Gerhard Svolba**

*Debugging SAS® Programs: A Handbook of Tools and
Techniques*
by **Michele M. Burlew**

*Decision Trees for Business Intelligence and Data Mining: Using
SAS® Enterprise Miner™*
by **Barry de Ville**

*Efficiency: Improving the Performance of Your SAS®
Applications*
by **Robert Virgile**

Elementary Statistics Using JMP®
by **Sandra D. Schlotzhauer**

The Essential Guide to SAS® Dates and Times
by **Derek P. Morgan**

The Essential PROC SQL Handbook for SAS® Users
by **Katherine Prairie**

*Fixed Effects Regression Methods for Longitudinal Data
Using SAS®*
by **Paul D. Allison**

Genetic Analysis of Complex Traits Using SAS®
Edited by **Arnold M. Saxton**

A Handbook of Statistical Analyses Using SAS®, Second Edition
by **B.S. Everitt**
and **G. Der**

Health Care Data and SAS®
by **Marge Scerbo, Craig Dickstein,**
and **Alan Wilson**

The How-To Book for SAS/GRAPH® Software
by **Thomas Miron**

*In the Know ... SAS® Tips and Techniques From
Around the Globe, Second Edition*
by **Phil Mason**

Instant ODS: Style Templates for the Output Delivery System
by **Bernadette Johnson**

*Integrating Results through Meta-Analytic Review Using
SAS® Software*
by **Morgan C. Wang**
and **Brad J. Bushman**

Introduction to Data Mining Using SAS® Enterprise Miner™
by **Patricia B. Cerrito**

Learning SAS® by Example: A Programmer's Guide
by **Ron Cody**

Learning SAS® in the Computer Lab, Second Edition
by **Rebecca J. Elliott**

support.sas.com/pubs

SAS® System for Regression, Third Edition
by **Rudolf J. Freund**
and **Ramon C. Littell**

SAS® System for Statistical Graphics, First Edition
by **Michael Friendly**

The SAS® Workbook and *Solutions* Set
(books in this set also sold separately)
by **Ron Cody**

*Selecting Statistical Techniques for Social Science Data:
A Guide for SAS® Users*
by **Frank M. Andrews, Laura Klem, Patrick M. O'Malley,
Willard L. Rodgers, Kathleen B. Welch,**
and **Terrence N. Davidson**

Statistical Quality Control Using the SAS® System
by **Dennis W. King**

Statistics Using SAS® Enterprise Guide®
by **James B. Davis**

*A Step-by-Step Approach to Using the SAS® System
for Factor Analysis and Structural Equation Modeling*
by **Larry Hatcher**

*A Step-by-Step Approach to Using SAS® for Univariate and
Multivariate Statistics, Second Edition*
by **Norm O'Rourke, Larry Hatcher,**
and **Edward J. Stepanski**

Step-by-Step Basic Statistics Using SAS®: Student Guide
and *Exercises*
*(*books in this set also sold separately)
by **Larry Hatcher**

*Survival Analysis Using SAS®:
A Practical Guide*
by **Paul D. Allison**

*Tuning SAS® Applications in the OS/390 and z/OS
Environments, Second Edition*
by **Michael A. Raithel**

*Univariate and Multivariate General Linear Models:
Theory and Applications Using SAS® Software*
by **Neil H. Timm**
and **Tammy A. Mieczkowski**

Using SAS® in Financial Research
by **Ekkehart Boehmer, John Paul Broussard,**
and **Juha-Pekka Kallunki**

Using the SAS® Windowing Environment: A Quick Tutorial
by **Larry Hatcher**

Visualizing Categorical Data
by **Michael Friendly**

Web Development with SAS® by Example, Second Edition
by **Frederick E. Pratter**

Your Guide to Survey Research Using the SAS® System
by **Archer Gravely**

JMP® Books

*JMP® for Basic Univariate and Multivariate Statistics: A Step-by-
Step Guide*
by **Ann Lehman, Norm O'Rourke, Larry Hatcher,**
and **Edward J. Stepanski**

JMP® Start Statistics, Third Edition
by **John Sall, Ann Lehman,**
and **Lee Creighton**

Regression Using JMP®
by **Rudolf J. Freund, Ramon C. Littell,**
and **Lee Creighton**